North Carolina State Building Code:
Energy Conservation Code

(2009 IECC® with North Carolina Amendments)

2012

2012 North Carolina Energy Conservation Code

First Printing: January 2012
Second Printing: March 2012
Third Printing: July 2012

ISBN-978-1-60983-122-6

NORTH CAROLINA STATE BUILDING CODE COUNCIL
MARCH 1, 2012
www.ncbuildingcodes.com

Al Bass, Jr., PE – 15
(Mechanical Engineer)
Bass, Nixon and Kennedy
6425 Chapman Court
Raleigh, NC 27612
919-782-4689

Cindy Browning, PE – 17
(State Agency)
State Construction
301 North Wilmington St.
Raleigh, NC 27601
919-807-4127

Ralph Euchner – 13
(Gas Industry)
PSNC Energy
PO Box 1398
Gastonia, NC 28053
704-810-3331

Leah Faile, AIA – 16
(Architect)
O'Brien/Atkins Assoc., PA
PO Box 12037
RTP, NC 27709
919-941-9000

Steve Knight, PE – 15
(Structural Engineer)
Steve L. Knight, PE
1507 Mount Vernon Ave.
Statesville, NC 28677
704-878-2996

Lon McSwain – 15
(Building Inspector)
Mecklenburg County
700 North Tryon Street
Charolotte, NC 28202
704-336-4302

CHAIR

Dan Tingen – 17
(Home Builder)
Tingen Construction Co.
8411 Garvey Drive, #101
Raleigh, NC 27616
919-875-2161

Ed Moore, Sr. – 13
(Electrical Contractor)
E. Moore and Son Electric
2708 N. Graham St., Ste. D
Charlotte, NC 28206
704-358-8828

Mack Nixon – 16
(County Representative)
Albemarle Home Builders
199 Mill Street
Elizabeth City, NC 27909
252-338-5211

Mack Paul – 17
(Public Representative)
K&L Gates, 4330 Lassiter
at North Hills, Ste. 300
Raleigh, NC 27101
919-743-7300

Alan Perdue – 15
(Fire Services)
Guilford County
1002 Meadowood Street
Greensboro, NC 27409
336-641-7565

Kim Reitterer, PE – 13
(Electrical Engineer)
ELM Engineering
212 S. Tryon St., Ste. 1050
Charlotte, NC 28281
704-335-0396

VICE CHAIRMAN

John Hitch, AIA – 16
(Architect)
The Smith Sinnett Assoc.
4601 Lake Boone Trail
Raleigh, NC 27607
919-781-8582

Bob Ruffner, Jr. – 15
(General Contractor)
Clancy and Theys Construction
PO Box 4189
Wilmington, NC 28406
910-392-5220

David Smith – 16
(Coastal Contractor)
D. Smith Builder
905 Saltwood Lane
Wilmington, NC 28411
910-681-0394

Scott Stevens – 16
(Municipal Representative)
City Manager
PO Drawer A
Goldsboro, NC 27533
919-580-4330

Paula Strickland – 13
(Mechanical Contractor)
Williams PH&AC
1051 Grecade Street
Greensboro, NC 27408
336-275-1328

NORTH CAROLINA
DEPARTMENT OF INSURANCE

www.ncdoi.com/osfm
919-661-5880

By Statute, the Commissioner of Insurance has general supervision of the administration and enforcement of the North Carolina State Building Code, and the Engineering Division serves as the Staff for the Building Code Council. Officials of the Department of Insurance are:

WAYNE GOODWIN
Commissioner

TIM BRADLEY
Senior Deputy Commissioner

CHRIS NOLES, PE
Deputy Commissioner

BARRY GUPTON, PE
Chief Code Consultant

DAN DITTMAN, PE
Energy Code Consultant

COMMITTEES OF THE COUNCIL
MARCH 1, 2012

ADMINISTRATION
Dan Tingen – Chair
Al Bass, PE
Ralph Euchner
John Hitch, AIA
Steve Knight, PE
Lon McSwain
Alan Perdue
Kim Reitterer, PE
David Smith

BUILDING
Lon McSwain – Chair
Cindy Browning, PE
Leah Faile, AIA
John Hitch, AIA
Ed Moore, Sr.
Alan Perdue
Bob Ruffner, Jr.
Paula Strickland

ELECTRICAL
Kim Reitterer, PE – Chair
Al Bass, PE
Cindy Browning, PE
Leah Faile, AIA
John Hitch, AIA
Ed Moore, Sr.
Bob Ruffner, Jr.

ENERGY
Ralph Euchner – Chair
Al Bass, PE
Leah Faile, AIA
Mack Nixon
Mack Paul
Kim Reitterer, PE
Bob Ruffner, Jr.
David Smith
Scott Stevens

FIRE PREVENTION
Alan Perdue – Chair
Ralph Euchner
John Hitch, AIA
Mack Nixon
Mack Paul
Bob Ruffner, Jr.
Scott Stevens

MECHANICAL
Al Bass, PE – Chair
Ralph Euchner
Ed Moore, Sr.
David Smith
Paula Strickland

RESIDENTIAL
David Smith – Chair
Cindy Browning, PE
Ralph Euchner
Steve Knight, PE
Lon McSwain
Mack Nixon
Scott Stevens
Paula Strickland

STRUCTURAL
Steve Knight, PE – Chair
Al Bass, PE
Leah Faile, AIA
John Hitch, AIA
Bob Ruffner, Jr.
Scott Stevens

PREFACE

Introduction

Internationally, code officials recognize the need for a modern, up-to-date energy conservation code addressing the design of energy-efficient building envelopes and installation of energy efficient mechanical, lighting and power systems through requirements emphasizing performance. The *International Energy Conservation Code®*, in this 2009 edition, is designed to meet these needs through model code regulations that will result in the optimal utilization of fossil fuel and nondepletable resources in all communities, large and small.

This comprehensive energy conservation code establishes minimum regulations for energy efficient buildings using prescriptive and performance-related provisions. It is founded on broad-based principles that make possible the use of new materials and new energy efficient designs. This 2009 edition is fully compatible with all the *International Codes®* (I-Codes®) published by the International Code Council (ICC)®, including: the *International Building Code®, International Existing Building Code®, International Fire Code®, International Fuel Gas Code®, International Mechanical Code®*, ICC *Performance Code®, International Plumbing Code®, International Private Sewage Disposal Code®, International Property Maintenance Code®, International Residential Code®, International Wildland-Urban Interface Code* and *International Zoning Code®*.

The *International Energy Conservation Code* provisions provide many benefits, among which is the model code development process that offers an international forum for energy professionals to discuss performance and prescriptive code requirements. This forum provides an excellent arena to debate proposed revisions. This model code also encourages international consistency in the application of provisions.

Development

The first edition of the *International Energy Conservation Code* (1998) was based on the 1995 edition of the *Model Energy Code* promulgated by the Council of American Building Officials (CABO) and included changes approved through the CABO Code Development Procedures through 1997. CABO assigned all rights and responsibilities to the International Code Council and its three statutory members at that time, including Building Officials and Code Administrators International, Inc. (BOCA), International Conference of Building Officials (ICBO) and Southern Building Code Congress International (SBCCI). This 2009 edition presents the code as originally issued, with changes reflected in the 2000, 2003 and 2006 editions and further changes approved through the ICC Code Development Process through 2008. A new edition such as this is promulgated every three years.

This code is founded on principles intended to establish provisions consistent with the scope of an energy conservation code that adequately conserves energy; provisions that do not unnecessarily increase construction costs; provisions that do not restrict the use of new materials, products or methods of construction; and provisions that do not give preferential treatment to particular types or classes of materials, products or methods of construction.

Adoption

The *International Energy Conservation Code* is available for adoption and use by jurisdictions internationally. Its use within a governmental jurisdiction is intended to be accomplished through adoption by reference in accordance with proceedings establishing the jurisdiction's laws. At the time of adoption, jurisdictions should insert the appropriate information in provisions requiring specific local information, such as the name of the adopting jurisdiction. These locations are shown in bracketed words in small capital letters in the code and in the sample ordinance. The sample adoption ordinance on page vii addresses several key elements of a code adoption ordinance, including the information required for insertion into the code text.

Maintenance

The *International Energy Conservation Code* is kept up to date through the review of proposed changes submitted by code enforcing officials, industry representatives, design professionals and other interested parties. Proposed changes are carefully considered through an open code development process in which all interested and affected parties may participate.

The contents of this work are subject to change both through the Code Development Cycles and the governmental body that enacts the code into law. For more information regarding the code development process, contact the Code and Standard Development Department of the International Code Council.

While the development procedure of the *International Energy Conservation Code* assures the highest degree of care, ICC, its members and those participating in the development of this code do not accept any liability resulting from compliance or noncompliance with the provisions because ICC and its members do not have the power or authority to police or enforce compliance with the contents of this code. Only the governmental body that enacts the code into law has such authority.

Marginal Markings

The 2012 *North Carolina Energy Conservation Code* is based on the requirements of the 2009 *International Energy Conservation Code*. Since there are numerous North Carolina amendments to that document, this code is printed without margin and underline markings.

Italicized Terms

Selected terms set forth in Chapter 2, Definitions, are italicized where they appear in code text. Such terms are not italicized where the definition set forth in Chapter 2 does not impart the intended meaning in the use of the term. The terms selected have definitions which the user should read carefully to facilitate better understanding of the code.

Effective Use of the North Carolina Energy Conservation Code

The *North Carolina Energy Conservation Code (NCECC)* is a model code that regulates minimum energy conservation requirements for new buildings. The NCECC addresses energy conservation requirements for all aspects of energy uses in both commercial and residential construction, including heating and ventilating, lighting, water heating, and power usage for appliances and building systems.

The NCECC is a design document. For example, before one constructs a building, the designer must determine the minimum insulation R-values and fenestration U-factors for the building exterior envelope. Depending on whether the building is for residential use or for commercial use, the NCECC sets forth minimum requirements for exterior envelope insulation, window and door U-factors and SHGC ratings, duct insulation, lighting and power efficiency, and water distribution insulation.

Arrangement and Format of the 2012 NCECC

Before applying the requirements of the NCECC it is beneficial to understand its arrangement and format. The NCECC is arranged and organized to follow sequential steps that generally occur during a plan review or inspection. The NCECC is divided into five different parts:

Chapters	Subjects
1–2	Administration and definitions
3	Climate zones and general materials requirements
4	Energy efficiency for residential buildings
5	Energy efficiency for commercial buildings
6	Referenced standards

The following is a chapter-by-chapter synopsis of the scope and intent of the provisions of the *NCEEC*.

Chapter 1 Administration. This chapter contains provisions for the application, enforcement and administration of subsequent requirements of the code. In addition to establishing the scope of the code, Chapter 1 identifies which buildings and structures come under its purview. Chapter 1 is largely concerned with maintaining "due process of law" in enforcing the energy conservation criteria contained in the body of the code. Only through careful observation of the administrative provisions can the building official reasonably expect to demonstrate that "equal protection under the law" has been provided.

Chapter 2 Definitions. All terms that are defined in the code are listed alphabetically in Chapter 2. While a defined term may be used in one chapter or another, the meaning provided in Chapter 2 is applicable throughout the code.

Additional definitions regarding climate zones are found in Tables 301.3(1) and (2). These are not listed in Chapter 2.

Where understanding of a term's definition is especially key to or necessary for understanding of a particular code provision, the term is shown in *italics* wherever it appears in the code. This is true only for those terms that have a meaning that is unique to the code. In other words, the generally understood meaning of a term or phrase might not be sufficient or consistent with the meaning prescribed by the code; therefore, it is essential that the code-defined meaning be known.

Guidance regarding tense, gender and plurality of defined terms as well as guidance regarding terms not defined in this code is provided.

Chapter 3 Climate Zones. Chapter 3 specifies the climate zones that will serve to establish the exterior design conditions. In addition, Chapter 3 provides interior design conditions that are used as a basis for assumptions in heating and cooling load calculations, and provides basic material requirements for insulation materials and fenestration materials.

Climate has a major impact on the energy use of most buildings. The code establishes many requirements such as wall and roof insulation R-values, window and door thermal transmittance requirement (U-factors) as well as provisions that affect the mechanical systems based upon the climate where the building is located. This chapter will contain the information that will be used to properly assign the building location into the correct climate zone and will then be used as the basis for establishing requirements or elimination of requirements.

Chapter 4 Residential Energy Efficiency. Chapter 4 contains the energy-efficiency-related requirements for the design and construction of residential buildings regulated under this code. It should be noted that the definition of a *residential building* in this code is unique for this code. In this code, a *residential building* is an R-2, R-3 or R-4 building three stories or less in height. All other R-1 buildings, including residential buildings greater than three stories in height, are regulated by the energy conservation requirements of Chapter 5. The applicable portions of a residential building must comply with the provisions within this chapter for energy efficiency. This chapter defines requirements for the portions of the building and building systems that impact energy use in new residential construction and promotes the effective use of energy. The provisions within the chapter promote energy efficiency in the building envelope, the heating and cooling system and the service water heating system of the building.

Chapter 5 Commercial Energy Efficiency. Chapter 5 contains the energy-efficiency-related requirements for the design and construction of most types of commercial buildings and residential buildings greater than three stories in height above grade. Residential buildings, townhouses and garden apartments three stories or less in height are covered in Chapter 4. Like Chapter 4, this chapter defines requirements for the portions of the building and building systems that impact energy use in new commercial construction and new residential construction greater than three stories in height, and promotes the effective use of energy. The provisions within the chapter promote energy efficiency in the building envelope, the heating and cooling system and the service water heating system of the building.

Chapter 6 Referenced Standards. The code contains numerous references to standards that are used to regulate materials and methods of construction. Chapter 6 contains a comprehensive list of all standards that are referenced in the code. The standards are part of the code to the extent of the reference to the standard. Compliance with the referenced standard is necessary for compliance with this code. By providing specifically adopted standards, the construction and installation requirements necessary for compliance with the code can be readily determined. The basis for code compliance is, therefore, established and available on an equal basis to the code official, contractor, designer and owner.

Chapter 6 is organized in a manner that makes it easy to locate specific standards. It lists all of the referenced standards, alphabetically, by acronym of the promulgating agency of the standard. Each agency's standards are then listed in either alphabetical or numeric order based upon the standard identification. The list also contains the title of the standard; the edition (date) of the standard referenced; any addenda included as part of the ICC adoption; and the section or sections of this code that reference the standard.

ORDINANCE

The International Codes are designed and promulgated to be adopted by reference by ordinance. Jurisdictions wishing to adopt the 2009 *International Energy Conservation Code* as an enforceable regulation governing energy efficient building envelopes and installation of energy efficient mechanical, lighting and power systems should ensure that certain factual information is included in the adopting ordinance at the time adoption is being considered by the appropriate governmental body. The following sample adoption ordinance addresses several key elements of a code adoption ordinance, including the information required for insertion into the code text.

SAMPLE ORDINANCE FOR ADOPTION OF THE *INTERNATIONAL ENERGY CONSERVATION CODE* ORDINANCE NO._____

An ordinance of the **[JURISDICTION]** adopting the 2009 edition of the *International Energy Conservation Code*, regulating and governing energy efficient building envelopes and installation of energy efficient mechanical, lighting and power systems in the **[JURISDICTION]**; providing for the issuance of permits and collection of fees therefor; repealing Ordinance No. _____ of the **[JURISDICTION]** and all other ordinances and parts of the ordinances in conflict therewith.

The **[GOVERNING BODY]** of the **[JURISDICTION]** does ordain as follows:

Section 1. That a certain document, three (3) copies of which are on file in the office of the **[TITLE OF JURISDICTION'S KEEPER OF RECORDS]** of **[NAME OF JURISDICTION]**, being marked and designated as the *International Energy Conservation Code*, 2009 edition, as published by the International Code Council, be and is hereby adopted as the Energy Conservation Code of the [JURISDIC-TION], in the State of [STATE NAME] for regulating and governing energy efficient building envelopes and installation of energy efficient mechanical, lighting and power systems as herein provided; providing for the issuance of permits and collection of fees therefor; and each and all of the regulations, provisions, penalties, conditions and terms of said Energy Conservation Code on file in the office of the **[JURISDICTION]** are hereby referred to, adopted, and made a part hereof, as if fully set out in this ordinance, with the additions, insertions, deletions and changes, if any, prescribed in Section 2 of this ordinance.

Section 2. The following sections are hereby revised:

Section 101.1. Insert: **[NAME OF JURISDICTION]**.

Section 108.4. Insert: **[DOLLAR AMOUNT]** in two places.

Section 3. That Ordinance No. _____ of **[JURISDICTION]** entitled **[FILL IN HERE THE COMPLETE TITLE OF THE ORDINANCE OR ORDI-NANCES IN EFFECT AT THE PRESENT TIME SO THAT THEY WILL BE REPEALED BY DEFINITE MENTION]** and all other ordinances or parts of ordinances in conflict herewith are hereby repealed.

Section 4. That if any section, subsection, sentence, clause or phrase of this ordinance is, for any reason, held to be unconstitutional, such decision shall not affect the validity of the remaining portions of this ordinance. The **[GOVERNING BODY]** hereby declares that it would have passed this ordinance, and each section, subsection, clause or phrase thereof, irrespective of the fact that any one or more sections, subsections, sentences, clauses and phrases be declared unconstitutional.

Section 5. That nothing in this ordinance or in the *International Energy Conservation Code*® hereby adopted shall be construed to affect any suit or proceeding impending in any court, or any rights acquired, or liability incurred, or any cause or causes of action acquired or existing, under any act or ordinance hereby repealed as cited in Section 3 of this ordinance; nor shall any just or legal right or remedy of any character be lost, impaired or affected by this ordinance.

Section 6. That the **[JURISDICTION'S KEEPER OF RECORDS]** is hereby ordered and directed to cause this ordinance to be published. (An additional provision may be required to direct the number of times the ordinance is to be published and to specify that it is to be in a newspaper in general circulation. Posting may also be required.)

Section 7. That this ordinance and the rules, regulations, provisions, requirements, orders and matters established and adopted hereby shall take effect and be in full force and effect **[TIME PERIOD]** from and after the date of its final passage and adoption.

TABLE OF CONTENTS

CHAPTER 1

ADMINISTRATION

PART 1—SCOPE AND APPLICATION

SECTION 101
SCOPE AND GENERAL REQUIREMENTS

101.1 Title. This code shall be known as the *North Carolina Energy Conservation Code* as adopted by the North Carolina Building Code Council on December 14, 2010, to be effective January 1, 2012. References to the International Codes shall mean the North Carolina Codes. The NCECC is referred to herein as "this code."

101.2 Scope. This code applies to *residential* and *commercial buildings*.

101.3 Intent. This code shall regulate the design and construction of buildings for the effective use of energy. This code is intended to provide flexibility to permit the use of innovative approaches and techniques to achieve the effective use of energy. This code is not intended to abridge safety, health or environmental requirements contained in other applicable codes or ordinances.

101.4 Applicability. Where, in any specific case, different sections of this code specify different materials, methods of construction or other requirements, the most restrictive shall govern. Where there is a conflict between a general requirement and a specific requirement, the specific requirement shall govern.

101.4.1 Existing buildings. Except as specified in this chapter, this code shall not be used to require the removal, *alteration* or abandonment of, nor prevent the continued use and maintenance of, an existing building or building system lawfully in existence at the time of adoption of this code.

101.4.2 Historic buildings. Any building or structure that is listed in the State or National Register of Historic Places; designated as a historic property under local or state designation law or survey; certified as a contributing resource with a National Register listed or locally designated historic district; or with an opinion or certification that the property is eligible to be listed on the National or State Registers of Historic Places either individually or as a contributing building to a historic district by the State Historic Preservation Officer or the Keeper of the National Register of Historic Places, are exempt from this code.

101.4.3 Additions, alterations, renovations or repairs. Additions, alterations, renovations or repairs to an existing building, building system or portion thereof shall conform to the provisions of this code as they relate to new construction without requiring the unaltered portion(s) of the existing building or building system to comply with this code. Additions, alterations, renovations or repairs shall not create an unsafe or hazardous condition or overload existing building systems. An addition shall be deemed to comply with this code if the addition alone complies or if the existing building and addition comply with this code as a single building.

Exceptions:

1. The following need not comply provided the energy use of the building is not increased:

    a. Storm windows installed over existing fenestration.

    b. Incidental repairs requiring a new sash or new glazing.

    c. Existing ceiling, wall or floor cavities exposed during construction provided that these cavities are filled with insulation.

    d. Construction where the existing roof, wall or floor cavity is not exposed.

    e. Reroofing for roofs where neither the sheathing nor the insulation is exposed. Roofs without insulation in the cavity and where the sheathing or insulation is exposed during reroofing shall be insulated either above or below the sheathing.

    f. Replacement of existing doors that separate *conditioned space* from the exterior shall not require the installation of a vestibule or revolving door, provided, however, that an existing vestibule that separates a *conditioned space* from the exterior shall not be removed,

    g. Alterations that replace less than 50 percent of the luminaires in a space, provided that such alterations do not increase the installed interior lighting power.

    h. Alterations that replace only the bulb and ballast within the existing luminaires in a space provided that the alteration does not increase the installed interior lighting power.

2. Converting unconditioned attic space to conditioned attic space for one and two-family dwellings and townhouses. Ceilings shall be insulated to a minimum of R-30, walls shall be insulated to the exterior wall requirements in Tables 402.1.1 and 402.1.3 and follow backing requirements in Section 402.2.12.

101.4.4 Change in occupancy or use. Spaces undergoing a change in occupancy that would result in an increase in demand for either fossil fuel or electrical energy shall comply with this code. Where the use in a space changes from one use in Table 505.5.2 to another use in Table 505.5.2, the installed lighting wattage shall comply with Section 505.5.

101.4.5 Change in space conditioning. Any nonconditioned space that is altered to become *conditioned space* shall be required to be brought into full compliance with this code.

Exception: See Section 101.4.3, exception 2.

101.4.6 Mixed occupancy. Where a building includes both *residential* and *commercial* occupancies, each occupancy shall be separately considered and meet the applicable provisions of Chapter 4 for *residential* and Chapter 5 for *commercial*.

101.5 Compliance. *Residential buildings* shall meet the provisions of Chapter 4. *Commercial buildings* shall meet the provisions of Chapter 5.

101.5.1 Compliance materials. The *code official* shall be permitted to approve specific computer software, worksheets, compliance manuals and other similar materials that meet the intent of this code.

101.5.2 Low energy buildings. The following buildings, or portions thereof, separated from the remainder of the building by *building thermal envelope* assemblies complying with this code shall be exempt from the *building thermal envelope* provisions of this code:

1. Those with a peak design rate of energy usage less than 3.4 Btu/h · ft^2 (10.7 W/m^2) or 1.0 watt/ft^2 (10.7 W/m^2) of floor area for space conditioning purposes.

2. Those that do not contain *conditioned space*.

101.5.3 Requirements of other State Agencies, occupational licensing boards, or commissions. The North Carolina State Building Codes do not include all additional requirements for buildings and structures that may be imposed by other State agencies, occupational licensing boards, and commissions. It shall be the responsibility of a permit holder, design professional, contractor, or occupational license holder to determine whether any additional requirements exist.

SECTION 102
ALTERNATE MATERIALS—
METHOD OF CONSTRUCTION,
DESIGN OR INSULATING SYSTEMS

102.1 General. This code is not intended to prevent the use of any material, method of construction, design or insulating system not specifically prescribed herein, provided that such construction, design or insulating system has been *approved* by the *code official* as meeting the intent of this code.

102.1.1 Above code programs. Deleted.

PART 2—ADMINISTRATION AND ENFORCEMENT

SECTION 103
CONSTRUCTION DOCUMENTS

103.1 General. Construction documents and other supporting data shall be submitted in one or more sets with each application for a permit. The construction documents shall be prepared by a registered design professional where required by the statutes of the jurisdiction in which the project is to be constructed.

Exceptions:

1. The *code official* is authorized to waive the requirements for construction documents or other supporting data if the *code official* determines they are not necessary to confirm compliance with this code.

2. Construction documents for energy code compliance are not required for one and two-family dwellings and townhouses.

103.2 Information on construction documents. Construction documents shall be drawn to scale. Electronic media documents are permitted to be submitted when *approved* by the *code official*. Construction documents shall be of sufficient clarity to indicate the location, nature and extent of the work proposed, and show in sufficient detail pertinent data and features of the building, systems and equipment as herein governed. Details shall include, as applicable, insulation materials and their *R*-values; fenestration *U*-factors and SHGCs; area-weighted *U*-factor and SHGC calculations; mechanical system design criteria; mechanical and service water heating system and equipment types, sizes and efficiencies; economizer description; equipment and systems controls; fan motor horsepower (hp) and controls; duct sealing, duct and pipe insulation and location; lighting fixture schedule with wattage and control narrative; and air sealing details.

103.3 Examination of documents. Deleted. See the *North Carolina Administrative Code and Policies*.

103.4 Amended construction documents. Deleted. See the *North Carolina Administrative Code and Policies*.

103.5 Retention of construction documents. Deleted. See the *North Carolina Administrative Code and Policies*.

SECTION 104
INSPECTIONS

104.1 General. Deleted. See the *North Carolina Administrative Code and Policies*.

SECTION 105
VALIDITY

105.1 General. Deleted.

SECTION 106
REFERENCED STANDARDS

106.1 General. The codes and standards referenced in this code shall be those listed in Chapter 6, and such codes and standards shall be considered as part of the requirements of this code to the prescribed extent of each such reference.

106.2 Conflicting requirements. Where the provisions of this code and the referenced standards conflict, the provisions of this code shall take precedence.

106.3 Application of references. References to chapter or section numbers, or to provisions not specifically identified by number, shall be construed to refer to such chapter, section or provision of this code.

106.4 Other laws. The provisions of this code shall not be deemed to nullify any provisions of local, state or federal law.

SECTION 107
FEES

Deleted. See the *North Carolina Administrative Code and Policies.*

SECTION 108
STOP WORK ORDER

Deleted. See the *North Carolina Administrative Code and Policies.*

SECTION 109
BOARD OF APPEALS

Deleted. See the *North Carolina Administrative Code and Policies.*

CHAPTER 2

DEFINITIONS

SECTION 201
GENERAL

201.1 Scope. Unless stated otherwise, the following words and terms in this code shall have the meanings indicated in this chapter.

201.2 Interchangeability. Words used in the present tense include the future; words in the masculine gender include the feminine and neuter; the singular number includes the plural and the plural includes the singular.

201.3 Terms defined in other codes. Terms that are not defined in this code but are defined in the *International Building Code, International Fire Code, International Fuel Gas Code, International Mechanical Code, International Plumbing Code* or the *International Residential Code* shall have the meanings ascribed to them in those codes.

201.4 Terms not defined. Terms not defined by this chapter shall have ordinarily accepted meanings such as the context implies.

SECTION 202
GENERAL DEFINITIONS

ABOVE-GRADE WALL. A wall more than 50 percent above grade and enclosing *conditioned space*. This includes between-floor spandrels, peripheral edges of floors, roof and basement knee walls, dormer walls, gable end walls, walls enclosing a mansard roof and skylight shafts.

ACCESSIBLE. Admitting close approach as a result of not being guarded by locked doors, elevation or other effective means (see "Readily *accessible*").

ACH50. Air Changes per Hour of measured air flow in relation to the building volume while the building is maintained at a pressure difference of 50 Pascals.

ADDITION. An extension or increase in the *conditioned space* floor area or height of a building or structure.

AIR BARRIER MATERIAL. Material(s) that have an air permeability not to exceed 0.004 cfm/ft² under a pressure differential of 0.3 in. water (1.57psf) (0.02 L/s.m² @ 75 Pa) when tested in accordance with ASTM E 2178.

AIR BARRIER SYSTEM. Material(s) assembled and joined together to provide a barrier to air leakage through the building envelope. An air barrier system is a combination of AIR BARRIER MATERIALS and sealants.

ALTERATION. Any construction or renovation to an existing structure other than repair or addition that requires a permit. Also, a change in a mechanical system that involves an extension, addition or change to the arrangement, type or purpose of the original installation that requires a permit.

APPROVED. Acceptable to the code official for compliance with the provisions of the applicable Code or reference standard.

AUTOMATIC. Self-acting, operating by its own mechanism when actuated by some impersonal influence, as, for example, a change in current strength, pressure, temperature or mechanical configuration (see "Manual").

BASEMENT WALL. A wall 50 percent or more below grade and enclosing *conditioned space*.

BPI ENVELOPE PROFESSIONAL. An individual that has passed the Building Performance Institute written and field examination requirements for the Building Envelope certification.

BUILDING. Any structure used or intended for supporting or sheltering any use or occupancy.

BUILDING COMMISSIONING AUTHORITY CERTIFIED COMMISSIONING PROFESSIONAL. An individual that has passed the eligibility and certification process maintained by the Building Commissioning Authority.

BUILDING THERMAL ENVELOPE. The basement walls, exterior walls, floor, roof, and any other building element that enclose *conditioned space*. This boundary also includes the boundary between *conditioned space* and any exempt or unconditioned space.

***C*-FACTOR (THERMAL CONDUCTANCE).** The coefficient of heat transmission (surface to surface) through a building component or assembly, equal to the time rate of heat flow per unit area and the unit temperature difference between the warm side and cold side surfaces (Btu/h × ft² × °F) [W/(m² × K)].

CFM25. Cubic Feet per minute of measured air flow while the forced air system is maintained at a pressure difference of 25 Pascals (0.1 inches w.p.).

CFM50. Cubic Feet per Minute of measured air flow while the building is maintained at a pressure difference of 50 Pascals (0.2 inches w.p.).

CLOSED CRAWL SPACE. A foundation without wall vents that uses air sealed walls, ground and foundation moisture control, and mechanical drying potential to control crawl space moisture. Insulation may be located at the floor level or at the exterior walls.

CODE OFFICIAL. The officer or other designated authority charged with the administration and enforcement of this code, or a duly authorized representative.

COMMERCIAL BUILDING. For this code, all buildings that are not included in the definition of "Residential buildings."

CONDITIONED CRAWL SPACE. A foundation without wall vents that encloses an intentionally heated or cooled space. Insulation is located at the exterior walls.

CONDITIONED FLOOR AREA. The horizontal projection of the floors associated with the *conditioned space.*

CONDITIONED SPACE. For energy purposes, space within a building that is provided with heating or cooling equipment or systems capable of maintaining, through design or heat loss/gain, 50°F (10°C) during the heating season and 85°F (29°C) during the cooling season, or communicates directly with a conditioned space. For mechanical purposes, an area, room or space being heated or cooled by any equipment or appliance.

CRAWL SPACE WALL. The opaque portion of a wall that encloses a crawl space and is partially or totally below grade.

CURTAIN WALL. Fenestration products used to create an external nonload-bearing wall that is designed to separate the exterior and interior environments.

DAYLIGHT ZONE.

1. **Under skylights.** The area under skylights whose horizontal dimension, in each direction, is equal to the skylight dimension in that direction plus either the floor-to-ceiling height or the dimension to a ceiling height opaque partition, or one-half the distance to adjacent skylights or vertical fenestration, whichever is least.

2. **Adjacent to vertical fenestration.** The area adjacent to vertical fenestration which receives daylight through the fenestration. For purposes of this definition and unless more detailed analysis is provided, the daylight zone depth is assumed to extend into the space a distance of 15 feet (4572 mm) or to the nearest ceiling height opaque partition, whichever is less. The daylight zone width is assumed to be the width of the window plus 2 feet (610 mm) on each side, or the window width plus the distance to an opaque partition, or the window width plus one-half the distance to adjacent skylight or vertical fenestration, whichever is least.

DEMAND CONTROL VENTILATION (DCV). A ventilation system capability that provides for the automatic reduction of outdoor air intake below design rates when the actual occupancy of spaces served by the system is less than design occupancy.

DUCT. A tube or conduit utilized for conveying air. The air passages of self-contained systems are not to be construed as air ducts.

DUCT SYSTEM. A continuous passageway for the transmission of air that, in addition to ducts, includes duct fittings, dampers, plenums, fans and accessory air-handling equipment and appliances.

DWELLING UNIT. A single unit providing complete independent living facilities for one or more persons, including permanent provisions for living, sleeping, eating, cooking and sanitation.

ECONOMIZER, AIR. A duct and damper arrangement and automatic control system that allows a cooling system to supply outside air to reduce or eliminate the need for mechanical cooling during mild or cold weather.

ECONOMIZER, WATER. A system where the supply air of a cooling system is cooled indirectly with water that is itself cooled by heat or mass transfer to the environment without the use of mechanical cooling.

ENERGY ANALYSIS. A method for estimating the annual energy use of the *proposed design* and *standard reference design* based on estimates of energy use.

ENERGY COST. The total estimated annual cost for purchased energy for the building functions regulated by this code, including applicable demand charges.

ENERGY RECOVERY VENTILATION SYSTEM. Systems that employ air-to-air heat exchangers to recover energy from exhaust air for the purpose of preheating, precooling, humidifying or dehumidifying outdoor ventilation air prior to supplying the air to a space, either directly or as part of an HVAC system.

ENERGY SIMULATION TOOL. An *approved* software program or calculation-based methodology that projects the annual energy use of a building.

ENTRANCE DOOR. Fenestration products used for ingress, egress and access in nonresidential buildings, including exterior entrances that utilize latching hardware and automatic closers and contain over 50-percent glass specifically designed to withstand heavy use and possibly abuse.

EXTERIOR WALL. Walls including both above-grade walls and basement walls.

FAN BRAKE HORSEPOWER (BHP). The horsepower delivered to the fan's shaft. Brake horsepower does not include the mechanical drive losses (belts, gears, etc.).

FAN SYSTEM BHP. The sum of the fan brake horsepower of all fans that are required to operate at fan system design conditions to supply air from the heating or cooling source to the *conditioned space(s)* and return it to the source or exhaust it to the outdoors.

FAN SYSTEM DESIGN CONDITIONS. Operating conditions that can be expected to occur during normal system operation that result in the highest supply fan airflow rate to conditioned spaces served by the system.

FAN SYSTEM MOTOR NAMEPLATE HP. The sum of the motor nameplate horsepower of all fans that are required to operate at design conditions to supply air from the heating or cooling source to the *conditioned space(s)* and return it to the source or exhaust it to the outdoors.

FENESTRATION. Skylights, roof windows, vertical windows (fixed or moveable), opaque doors, glazed doors, glazed block and combination opaque/glazed doors. Fenestration includes products with glass and nonglass glazing materials.

F-**FACTOR.** The perimeter heat loss factor for slab-on-grade floors (Btu/h × ft × °F) [W/(m × K)].

FULLY ENCLOSED ATTIC FLOOR SYSTEM. The ceiling insulation is enclosed on all six sides by an air barrier system, such as taped drywall below, solid framing joists on the sides, solid blocking on the ends, and solid sheathing on top which totally enclose the insulation. This system provides for full depth insulation over the exterior walls.

FULLY SHIELDED. A light fixture constructed, installed, and maintained in such a manner that all light emitted from the fixture, either directly from the lamp or a diffusing element, or indirectly by reflection or refraction from any part of the fixture, is projected below the horizontal plane through the fixture's lowest light emitting part.

HEAT TRAP. An arrangement of piping and fittings, such as elbows or a commercially available heat trap that prevents thermosiphoning of hot water during standby periods.

HEATED SLAB. Slab-on-grade construction in which the heating elements, hydronic tubing, or hot air distribution system is in contact with, or placed within or under, the slab.

HERS RATER. An individual that has completed training and been certified by RESNET (Residential Energy Services Network) Accredited Rating Provider.

HIGH-EFFICACY LAMPS. Compact fluorescent lamps, T-8 or smaller diameter linear fluorescent lamps, or lamps with a minimum efficacy of:

1. 60 lumens per watt for lamps over 40 watts,

2. 50 lumens per watt for lamps over 15 watts to 40 watts, and

3. 40 lumens per watt for lamps 15 watts or less.

HUMIDISTAT. A regulatory device, actuated by changes in humidity, used for automatic control of relative humidity.

INFILTRATION. The uncontrolled inward air leakage into a building caused by the pressure effects of wind or the effect of differences in the indoor and outdoor air density or both.

INSULATING SHEATHING. An insulating board with a core material having a minimum *R*-value of R-2.

LABELED. Appliance, equipment, materials or products to which have been affixed a label, seal, symbol or other identifying mark of a nationally recognized testing laboratory, inspection agency or other organization concerned with product evaluation that maintains periodic inspection of the production of the above-labeled items and whose labeling indicates either that the appliance, equipment, material or product meets identified standards or has been tested and found suitable for a specified purpose. *(Laboratories, agencies or organizations that have been identified by approval and accreditation bodies, such as ANSI, IAS, ICC or OSHA, are acceptable.)*

LAMP. The device in a lighting fixture that provides illumination, typically a bulb, fluorescent tube, or light emitting diode (LED).

LISTED. Appliance, equipment, materials, products or services included in a list published by an organization approved by the *code official* and concerned with evaluation of products or services that maintains periodic inspection of production of *listed* appliance, equipment or materials or periodic evaluation of services and whose listing states either that the equipment, material, product or service meets identified standards or has been tested and found suitable for a specified purpose.

LOW-VOLTAGE LIGHTING. Lighting equipment powered through a transformer such as a cable conductor, a rail conductor and track lighting.

MASS WALL. Masonry or concrete walls having a mass greater than or equal to 30 pounds per square foot (146 kg/m²). solid wood walls having a mass greater than 20 pounds per square foot (98 kg/m²), and any other walls having a heat capacity greater than or equal to 6 Btu/ft² ·°F[266 J/(m² · K)].

MANUAL. Capable of being operated by personal intervention (see "Automatic").

NAMEPLATE HORSEPOWER. The nominal motor horsepower rating stamped on the motor nameplate.

ON-SITE RENEWABLE ENERGY. Includes solar photovoltaic; active solar thermal that employs collection panels, heat transfer mechanical components; wind; small hydro; tidal; wave energy; geothermal (core earth); biomass energy systems; landfill gas and bio-fuel based electrical production, Onsite energy shall be generated on or adjacent to the project site and shall not be delivered to the project through the utility service.

PROCESS ENERGY. Energy consumed in support of manufacturing, industrial, or commercial process other than conditioning spaces and maintaining comfort and amenities for the occupants of a building.

PROPOSED DESIGN. A description of the proposed building used to estimate annual energy use for determining compliance based on total building performance.

READILY ACCESSIBLE. Capable of being reached quickly for operation, renewal or inspection without requiring those to whom ready access is requisite to climb over or remove obstacles or to resort to portable ladders or access equipment (see "*Accessible*").

REGISTERED DESIGN PROFESSIONAL. An individual who is registered or licensed to practice his respective design profession as defined by the statutory requirements of the professional registration laws of the state or jurisdiction in which the project is to be constructed. Design by a Registered Design Professional is not required where exempt under the registration or licensure laws.

REPAIR. The reconstruction or renewal of any part of an existing building.

RESIDENTIAL BUILDING. For this code, includes R-3 buildings, as well as R-2 and R-4 buildings three stories or less in height above grade.

ROOF ASSEMBLY. A system designed to provide weather protection and resistance to design loads. The system consists of a roof covering and roof deck or a single component serving as both the roof covering and the roof deck. A roof assembly includes the roof covering, underlayment, roof deck, insulation, vapor retarder and interior finish.

***R*-VALUE (THERMAL RESISTANCE).** The inverse of the time rate of heat flow through a body from one of its bounding surfaces to the other surface for a unit temperature difference between the two surfaces, under steady state conditions, per unit area (h × ft² × °F/Btu) [(m² × K)/W].

SCREW LAMP HOLDERS. A lamp base that requires a screw-in-type lamp, such as a compact-fluorescent, incandescent, or tungsten-halogen bulb.

SEMI-CONDITIONED SPACE. A space indirectly conditioned within the thermal envelope that is not directly heated or cooled. For energy purposes, semi-conditioned spaces are treated as conditioned spaces.

SERVICE WATER HEATING. Supply of hot water for purposes other than comfort heating.

SHADING COEFFICIENT. The amount of the sun's heat transmitted through a given window compared with that of a standard 1/8- inch-thick single pane of glass under the same conditions.

SKYLIGHT. Glass or other transparent or translucent glazing material installed at a slope of 15 degrees (0.26 rad) or more from vertical. Glazing material in skylights, including unit skylights, solariums, sunrooms, roofs and sloped walls is included in this definition.

SLEEPING UNIT. A room or space in which people sleep, which can also include permanent provisions for living, eating, and either sanitation or kitchen facilities but not both. Such rooms and spaces that are also part of a dwelling unit are not sleeping units.

SOLAR HEAT GAIN COEFFICIENT (SHGC). The ratio of the solar heat gain entering the space through the fenestration assembly to the incident solar radiation. Solar heat gain includes directly transmitted solar heat and absorbed solar radiation which is then reradiated, conducted or convected into the space. This value is related to the Shading Coefficient (SC) by the formula SHGC = 0.87 · SC.

STANDARD REFERENCE DESIGN. A version of the *proposed design* that meets the minimum requirements of this code and is used to determine the maximum annual energy use requirement for compliance based on total building performance.

STOREFRONT. A nonresidential system of doors and windows mulled as a composite fenestration structure that has been designed to resist heavy use. *Storefront* systems include exterior fenestration systems that span from the floor level or above to the ceiling of the same story on commercial buildings.

SUNROOM. A one-story structure attached to a dwelling with a glazing area in excess of 40 percent of the gross area of the structure's exterior walls and roof.

SYSTEM VERIFICATION. A process that verifies and documents that the selected building systems have been designed, installed, and function according to the code requirements and construction documents.

THERMAL ISOLATION. Physical and space conditioning separation from *conditioned space(s)*. The *conditioned space(s)* shall be controlled as separate zones for heating and cooling or conditioned by separate equipment.

THERMOSTAT. An automatic control device used to maintain temperature at a fixed or adjustable set point.

U-**FACTOR (THERMAL TRANSMITTANCE).** The coefficient of heat transmission (air to air) through a building component or assembly, equal to the time rate of heat flow per unit area and unit temperature difference between the warm side and cold side air films (Btu/h × ft² × °F) [W/(m² × K)].

VENTILATION. The natural or mechanical process of supplying conditioned or unconditioned air to, or removing such air from, any space.

VENTILATION AIR. That portion of supply air that comes from outside (outdoors) plus any recirculated air that has been treated to maintain the desired quality of air within a designated space.

WALL VENTED CRAWL SPACE. A foundation that uses foundation wall vents as a primary means to control space moisture. Insulation is located at the floor level.

ZONE. A space or group of spaces within a building with heating or cooling requirements that are sufficiently similar so that desired conditions can be maintained throughout using a single controlling device.

CHAPTER 3
GENERAL REQUIREMENTS

SECTION 301
CLIMATE ZONES

301.1 General. Climate *zones* from Figure 301.1 or Table 301.1 shall be used in determining the applicable requirements from Chapters 4 and 5.

301.2 Warm humid counties. Warm humid counties are identified in Table 301.1 by an asterisk.

301.3 International climate zones. Deleted.

TABLE 301.1
NORTH CAROLINA CLIMATE ZONES, MOISTURE REGIMES, AND WARM-HUMID DESIGNATIONS BY COUNTY

Key: A – Moist, B – Dry, C – Marine. Absence of moisture designation indicates moisture regime is irrelevant. Asterisk (*) indicates a warm-humid location.

NORTH CAROLINA

4A Alamance	4A Cleveland	4A Henderson	3A Onslow*	3A Tyrrell
4A Alexander	3A Columbus*	4A Hertford	4A Orange	3A Union
5A Alleghany	3A Craven	3A Hoke	3A Pamlico	4A Vance
3A Anson	3A Cumberland	3A Hyde	3A Pasquotank	4A Wake
5A Ashe	3A Currituck	4A Iredell	3A Pender*	4A Warren
5A Avery	3A Dare	4A Jackson	3A Perquimans	3A Washington
3A Beaufort	3A Davidson	3A Johnston	4A Person	5A Watauga
4A Bertie	4A Davie	3A Jones	3A Pitt	3A Wayne
3A Bladen	3A Duplin	4A Lee	4A Polk	4A Wilkes
3A Brunswick*	4A Durham	3A Lenoir	3A Randolph	3A Wilson
4A Buncombe	3A Edgecombe	4A Lincoln	3A Richmond	4A Yadkin
4A Burke	4A Forsyth	4A Macon	3A Robeson	5A Yancey
3A Cabarrus	4A Franklin	4A Madison	4A Rockingham	
4A Caldwell	3A Gaston	3A Martin	3A Rowan	
3A Camden	4A Gates	4A McDowell	4A Rutherford	
3A Carteret*	4A Graham	3A Mecklenburg	3A Sampson	
4A Caswell	4A Granville	5A Mitchell	3A Scotland	
4A Catawba	3A Greene	3A Montgomery	3A Stanly	
4A Chatham	4A Guilford	3A Moore	4A Stokes	
4A Cherokee	4A Halifax	4A Nash	4A Surry	
3A Chowan	4A Harnett	3A New Hanover*	4A Swain	
4A Clay	4A Haywood	4A Northampton	4A Transylvania	

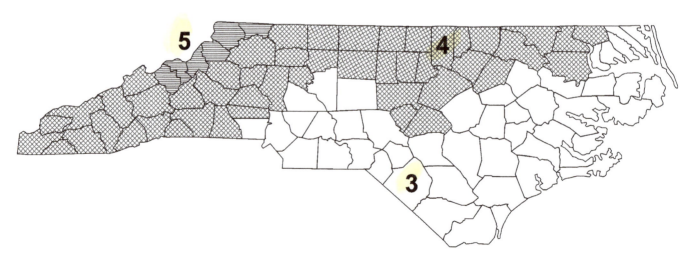

FIGURE 301.1
NORTH CAROLINA CLIMATE ZONES

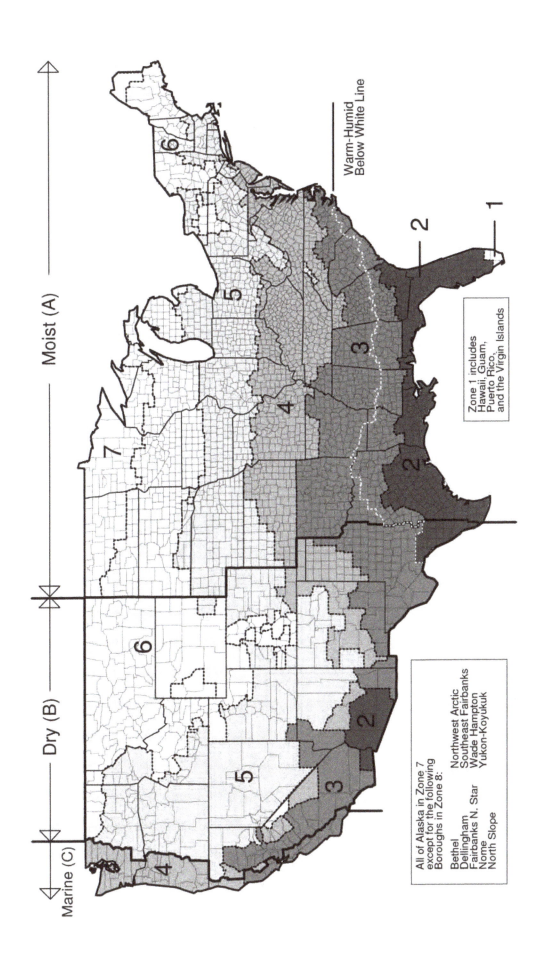

Moist (A)

Dry (B)

Marine (C)

Warm-Humid
Below White Line

Zone 1 includes
Hawaii, Guam,
Puerto Rico,
and the Virgin Islands

All of Alaska in Zone 7
except for the following
Boroughs in Zone 8:

Bethel   Northwest Arctic
Dellingham   Southeast Fairbanks
Fairbanks N. Star Wade Hampton
Nome   Yukon-Koyukuk
North Slope

FIGURE 301.2
CLIMATE ZONES

TABLE 301.3(1)
INTERNATIONAL CLIMATE ZONE DEFINITIONS

MAJOR CLIMATE TYPE DEFINITIONS
Warm-humid Definition—Moist (A) locations where either of the following wet-bulb temperature conditions shall occur during the warmest six consecutive months of the year: 1. 67°F (19.4°C) or higher for 3,000 or more hours; or 2. 73°F (22.8°C) or higher for 1,500 or more hours
Dry (B) Definition—Locations meeting the following criteria: Not marine and $P_{in} < 0.44 \times (TF - 19.5)$ $[P_{cm} < 2.0 \times (TC + 7)$ in SI units] where: P_{in} = Annual precipitation in inches (cm) T = Annual mean temperature in °F (°C)
Moist (A) Definition—Locations that are not marine and not dry.

For SI: °C = [(°F)-32]/1.8; 1 inch = 2.54 cm.

TABLE 301.3(2)
INTERNATIONAL CLIMATE ZONE DEFINITIONS

ZONE NUMBER	THERMAL CRITERIA	
	IP Units	SI Units
1	9000 < CDD50°F	5000 < CDD10°C
2	6300 < CDD50°F ≤ 9000	3500 < CDD10°C ≤ 5000
3A and 3B	4500 < CDD50°F ≤ 6300 AND HDD65°F ≤ 5400	2500 < CDD10°C ≤ 3500 AND HDD18°C ≤ 3000
4A and 4B	CDD50°F ≤ 4500 AND HDD65°F ≤ 5400	CDD10°C ≤ 2500 AND HDD18°C ≤ 3000
3C	HDD65°F ≤ 3600	HDD18°C ≤ 2000
4C	3600 < HDD65°F ≤ 5400	2000 < HDD18°C ≤ 3000
5	5400 < HDD65°F ≤ 7200	3000 < HDD18°C ≤ 4000
6	7200 < HDD65°F ≤ 9000	4000 < HDD18°C ≤ 5000
7	9000 < HDD65°F ≤ 12600	5000 < HDD18°C ≤ 7000
8	12600 < HDD65°F	7000 < HDD18°C

For SI: °C = [(°F)-32]/1.8

SECTION 302
DESIGN CONDITIONS

302.1 Interior design conditions. The interior design temperatures used for heating and cooling load calculations shall be a maximum of 72°F (22°C) for heating and minimum of 75°F (24°C) for cooling.

SECTION 303
MATERIALS, SYSTEMS AND EQUIPMENT

303.1 Identification. Materials, systems and equipment shall be identified in a manner that will allow a determination of compliance with the applicable provisions of this code.

303.1.1 Building thermal envelope insulation. An *R*-value identification mark shall be applied by the manufacturer to each piece of *building thermal envelope* insulation 12 inches (305 mm) or greater in width. Alternately, the insulation installers shall provide a certification listing the type, manufacturer and *R*-value of insulation installed in each element of the *building thermal envelope*. For blown or sprayed insulation (fiberglass and cellulose), the initial installed thickness, settled thickness, settled *R*-value, installed density, coverage area and number of bags installed shall be *listed* on the certification. For sprayed polyurethane foam (SPF) insulation, the installed thickness of the areas covered and *R*-value of installed thickness shall be *listed* on the certification. The insulation installer shall sign, date and post the certification in a conspicuous location on the job site.

303.1.1.1 Blown or sprayed roof/ceiling insulation. The thickness of blown-in or sprayed roof/ceiling insulation (fiberglass or cellulose) shall be written in inches (mm) on markers that are installed at least one for every 300 square feet (28 m²) throughout the attic space. The markers shall be affixed to the trusses or joists and marked with the minimum initial installed thickness with numbers a minimum of 1 inch (25 mm) in height. Each marker shall face the attic access opening. Spray polyurethane foam thickness and installed *R*-value shall be *listed* on certification provided by the insulation installer.

303.1.2 Insulation mark installation. Insulating materials shall be installed such that the manufacturer's *R*-value mark is readily observable upon inspection.

303.1.3 Fenestration product rating. *U*-factors of fenestration products (windows, doors and skylights) shall be determined in accordance with NFRC 100 by an accredited (by a body such as IAS or NFRC), independent laboratory, and labeled and certified by the manufacturer. Products lacking such a labeled *U*-factor shall be assigned a default *U*-factor from Table 303.1.3(1) or 303.1.3(2). The solar heat gain coefficient (SHGC) of glazed fenestration products (windows, glazed doors and skylights) shall be determined in accordance with NFRC 200 by an accredited, independent laboratory, and labeled and certified by the manufacturer. Products lacking such a labeled SHGC shall be assigned a default SHGC from Table 303.1.3(3).

TABLE 303.1.3(1)
DEFAULT GLAZED FENESTRATION *U*-FACTOR

FRAME TYPE	SINGLE PANE	DOUBLE PANE	SKYLIGHT	
			Single	Double
Metal	1.20	0.80	2.00	1.30
Metal with Thermal Break	1.10	0.65	1.90	1.10
Nonmetal or Metal Clad	0.95	0.55	1.75	1.05
Glazed Block	0.60			

TABLE 303.1.3(2)
DEFAULT DOOR *U*-FACTORS

DOOR TYPE	*U*-FACTOR
Uninsulated Metal	1.20
Insulated Metal	0.60
Wood	0.50
Insulated, nonmetal edge, max 45% glazing, any glazing double pane	0.35

TABLE 303.1.3(3)
DEFAULT GLAZED FENESTRATION SHGC

SINGLE GLAZED		DOUBLE GLAZED		
Clear	Tinted	Clear	Tinted	GLAZED BLOCK
0.8	0.7	0.7	0.6	0.6

303.1.4 Insulation product rating. The thermal resistance (*R*-value) of insulation shall be determined in accordance with the U.S. Federal Trade Commission *R*-value rule (CFR Title 16, Part 460, May 31, 2005) in units of h × ft² × °F/Btu at a mean temperature of 75°F (24°C).

303.2 Installation. All materials, systems and equipment shall be installed in accordance with the manufacturer's installation instructions and the *International Building Code.*

303.2.1 Protection of exposed foundation insulation. Insulation applied to the exterior of basement walls, crawlspace walls and the perimeter of slab-on-grade floors shall have a rigid, opaque and weather-resistant protective covering to prevent the degradation of the insulation's thermal performance. The protective covering shall cover the exposed exterior insulation and extend a minimum of 6 inches (153 mm) below grade.

303.3 Maintenance information. Maintenance instructions shall be furnished for equipment and systems that require preventive maintenance. Required regular maintenance actions shall be stated and incorporated on a readily accessible label. The label shall include the title or publication number for the operation and maintenance manual for that particular model and type of product.

CHAPTER 4

RESIDENTIAL ENERGY EFFICIENCY

SECTION 401
GENERAL

401.1 Scope. This chapter applies to residential buildings.

401.2 Compliance. Projects shall comply with Sections 401, 402.4, 402.5, and 403.1, 403.2.2, 403.2.3, and 403.3 through 403.9 (referred to as the mandatory provisions) and either:

1. Sections 402.1 through 402.3, 403.2.1 and 404.1 (prescriptive);

2. Section 405 (performance); or

3. North Carolina specific REScheck shall be permitted to demonstrate compliance with this code. Envelope requirements may not be traded off against the use of high efficiency heating or cooling equipment. No trade-off calculations are needed for required termite inspection and treatment gaps.

401.3 Certificate. A permanent certificate shall be posted on or in the electrical distribution panel, in the attic next to the attic insulation card, or inside a kitchen cabinet or other approved location. The certificate shall not cover or obstruct the visibility of the circuit directory label, service disconnect label or other required labels. The builder, permit holder, or registered design professional shall be responsible for completing the certificate.

The certificate shall list the predominant *R*-values of insulation installed in or on ceiling/roof, walls, foundation (slab, *basement wall*, crawlspace wall and floor) and ducts outside conditioned spaces; *U*-factors for fenestration and the solar heat gain coefficient (SHGC) of fenestration. Where there is more than one value for each component, the certificate shall list the value covering the largest area. The certificate shall indicate whether the building air leakage was visually inspected as required in 402.4.2.1 or provide results of the air leakage testing required in 402.4.2.2.The certificate shall provide results of duct leakage test required in Section 403.2.2. Appendix 1.1 contains a sample certificate.

401.4 Additional voluntary criteria for increasing residential energy efficiency. Appendix 4 contains additional voluntary measures for increasing residential energy efficiency beyond code minimums. Implementation of the increased energy efficiency measures is voluntary at the option of the permit holder. The sole purpose of the appendix is to provide guidance for achieving additional residential energy efficiency improvements that have been evaluated to be those that are most cost effective for achieving an additional 15-20% improvement in energy efficiency beyond code minimums.

TABLE 402.1.1
INSULATION AND FENESTRATION REQUIREMENTS BY COMPONENT[a]

CLIMATE ZONE	FENESTRATION *U*-FACTOR[b]	SKYLIGHT[b] *U*-FACTOR	GLAZED FENESTRATION SHGC[b, e]	CEILING[k] *R*-VALUE	WOOD FRAME WALL *R*-VALUE	MASS WALL *R*-VALUE[i]	FLOOR *R*-VALUE	BASEMENT[c] WALL *R*-VALUE	SLAB[d] *R*-VALUE & DEPTH	CRAWL SPACE[c] WALL *R*-VALUE
3	0.35	0.65	0.30	30	13	5/10	19	10/13[f]	0	5/13
4	0.35	0.60	0.30	38 or 30 cont.[j]	15, 13+2.5[h]	5/10	19	10/13	10	10/13
5	0.35	0.60	NR	38 or 30 cont.[j]	19, 13+5, or 15 + 3[e,h]	13/17	30[g]	10/13	10	10/13

For SI: 1 foot = 304.8 mm.

a. *R*-values are minimums. *U*-factors and SHGC are maximums.

b. The fenestration *U*-factor column excludes skylights. The SHGC column applies to all glazed fenestration.

c. "10/13" means R-10 continuous insulated sheathing on the interior or exterior of the home or R-13 cavity insulation at the interior of the basement wall or crawl space wall.

d. For monolithic slabs, insulation shall be applied from the inspection gap downward to the bottom of the footing or a maximum of 18 inches below grade whichever is less . For floating slabs, insulation shall extend to the bottom of the foundation wall or 24 inches, whichever is less. R-5 shall be added to the required slab edge *R*-values for heated slabs.

e. R-19 fiberglass batts compressed and installed in a nominal 2 × 6 framing cavity is deemed to comply. Fiberglass batts rated R-19 or higher compressed and installed in a 2 × 4 wall is not deemed to comply.

f. Basement wall insulation is not required in warm-humid locations as defined by Figure N1101.2(1 and 2) and Table N1101.2.

g. Or insulation sufficient to fill the framing cavity, R-19 minimum.

h. "13+5" means R-13 cavity insulation plus R-5 insulated sheathing. 15+3 means R-15 cavity insulation plus R-3 insulated sheathing. If structural sheathing covers 25 percent or less of the exterior, insulating sheathing is not required where structural sheathing is used. If structural sheathing covers more than 25 percent of exterior, structural sheathing shall be supplemented with insulated sheathing of at least R-2. 13+2.5 means R-13 cavity insulation plus R-2.5 sheathing.

i. For Mass Walls, the second R-value applies when more than half the insulation is on the interior of the mass wall.

j. R-30 shall be deemed to satisfy the ceiling insulation requirement wherever the full height of uncompressed R-30 insulation extends over the wall top plate at the eaves. Otherwise R-38 insulation is required where adequate clearance exists or insulation must extend to either the insulation baffle or within 1" of the attic roof deck.

k. Table value required except for roof edge where the space is limited by the pitch of the roof, there the insulation must fill the space up to the air baffle.

TABLE 402.1.3
EQUIVALENT *U*-FACTORS[a]

CLIMATE ZONE	FENESTRATION *U*-FACTOR	SKYLIGHT *U*-FACTOR	CEILING *U*-FACTOR	FRAME WALL *U*-FACTOR	MASS WALL *U*-FACTOR[b]	FLOOR *U*-FACTOR	BASEMENT WALL *U*-FACTOR	CRAWL SPACE WALL *U*-FACTOR[c]
3	0.35	0.65	0.035	0.082	0.141	0.047	0.059	0.136
4	0.35	0.60	0.030	0.077	0.141	0.047	0.059	0.065
5	0.35	0.60	0.030	0.061	0.082	0.033	0.059	0.065

a. Nonfenestration *U*-factors shall be obtained from measurement, calculation or an approved source.

b. When more than half the insulation is on the interior, the mass wall U-factors shall be a maximum of 0.17 in Zone 1, 0.14 in Zone 2, 0.12 in Zone 3, 0.10 in Zone 4 except Marine, and the same as the frame wall U-factor in Marine Zone 4 and Zones 5 through 8.

c. Basement wall *U*-factor of 0.360 in warm-humid locations as defined by Table 301.1 and Figure 301.2.

d. Foundation *U*-factor requirements shown in Table 402.1.3 include wall construction and interior air films but exclude soil conductivity and exterior air films. *U*-factors for determining code compliance in accordance with Section 402.1.4 (total UA alternative) of Section 405 (Simulated Performance Alternative) shall be modified to include soil conductivity and exterior air films.

SECTION 402
BUILDING THERMAL ENVELOPE

402.1 General (Prescriptive).

402.1.1 Insulation and fenestration criteria. The *building thermal envelope* shall meet the requirements of Table 402.1.1 based on the climate *zone* specified in Chapter 3.

402.1.2 *R*-value computation. Insulation material used in layers, such as framing cavity insulation and insulating sheathing, shall be summed to compute the component *R*-value. The manufacturer's settled *R*-value shall be used for blown insulation. Computed *R*-values shall not include an *R*-value for other building materials or air films.

402.1.3 *U*-factor alternative. An assembly with a *U*-factor equal to or less than that specified in Table 402.1.3 shall be permitted as an alternative to the *R*-value in Table 402.1.1.

402.1.4 Total UA alternative. If the total *building thermal envelope* UA (sum of *U*-factor times assembly area) is less than or equal to the total UA resulting from using the *U*-factors in Table 402.1.3 (multiplied by the same assembly area as in the proposed building), the building shall be considered in compliance with Table 402.1.1. The UA calculation shall be done using a method consistent with the ASHRAE *Handbook of Fundamentals* and shall include the thermal bridging effects of framing materials. The SHGC requirements shall be met in addition to UA compliance.

402.2 Specific insulation requirements (Prescriptive).

402.2.1 Ceilings with attic spaces. Ceilings with attic spaces over conditioned space shall meet the insulation requirements in Table 402.1.1.

Exceptions:

1. When insulation is installed in a fully *enclosed attic floor system,* as described in Appendix 1.2.1, R-30 shall be deemed compliant.

2. In roof edge and other details such as bay windows, dormers, and similar areas where the space is limited, the insulation must fill the space up to the air baffle.

402.2.2 Ceilings without attic spaces. Where the design of the roof/ceiling assembly, including cathedral ceilings, bay windows and other similar areas, does not allow sufficient space for the required insulation, the minimum required insulation for such roof/ceiling assemblies shall be R-30. This reduction of insulation from the requirements of Section 402.1.1 shall be limited to 500 square feet (46 m²) of ceiling surface area. This reduction shall not apply to the *U*-factor alternative approach in Section 402.1.3 and the total UA alternative in Section 402.1.4.

402.2.3 Access hatches and doors. Horizontal access hatches from conditioned spaces to unconditioned spaces (e.g., attics and crawl spaces) shall be weatherstripped and insulated to an R-10 minimum value, and vertical doors to such spaces shall be weatherstripped and insulated to R-5. Access shall be provided to all equipment that prevents damaging or compressing the insulation. A wood framed or equivalent baffle or retainer is required to be provided when loose fill insulation is installed, the purpose of which is to prevent the loose fill insulation from spilling into the living space when the attic access is opened, and to provide a permanent means of maintaining the installed *R*-value of the loose fill insulation.

Exceptions:

1. Pull down stair systems shall be weatherstripped and insulated to an R-5 insulation value such that the insulation does not interfere with proper operation of the stair. Non-rigid insulation materials are not allowed. Additional insulation systems that enclose the stair system from above are allowed. Exposed foam plastic must meet the provisions of the *Building Code* or *Residential Code,* respectively.

2. Full size doors that are part of the building thermal envelope and provide a passageway to unconditioned spaces shall meet the requirements of exterior doors in Section 402.3.4.

402.2.4 Mass walls. Mass walls for the purposes of this chapter shall be considered above-grade walls of concrete block, concrete, insulated concrete form (ICF), masonry cavity, brick (other than brick veneer), earth (adobe, compressed earth block, rammed earth) and solid timber/logs.

402.2.5 Steel-frame ceilings, walls, and floors. Steel-frame ceilings, walls and floors shall meet the insulation requirements of Table 402.2.5 or shall meet the *U*-factor requirements in Table 402.1.3. The calculation of the *U*-factor for a steel-frame envelope assembly shall use a series-parallel path calculation method.

TABLE 402.2.5
STEEL-FRAME CEILING, WALL AND FLOOR INSULATION
(*R*-VALUE)

WOOD FRAME *R*-VALUE REQUIREMENT	COLD-FORMED STEEL EQUIVALENT *R*-VALUE[a]
Steel Truss Ceilings[b]	
R-30	R-38 or R-30 + 3 or R-26 + 5
R-38	R-49 or R-38 + 3
R-49	R-38 + 5
Steel Joist Ceilings[b]	
R-30	R-38 in 2 × 4 or 2 × 6 or 2 × 8 R-49 in any framing
R-38	R-49 in 2 × 4 or 2 × 6 or 2 × 8 or 2 ×10
Steel-Framed Wall	
R-13	R-13 + 5 or R-15 + 4 or R-21 + 3 or R-0 +10
R-19	R-13 + 9 or R-19 + 8 or R-25 + 7
R-21	R-13 + 10 or R-19 + 9 or R-25 + 8
Steel Joist Floor	
R-13	R-19 in 2 × 6 R-19 + 6 in 2 × 8 or 2 ×10
R-19	R-19 + 6 in 2 × 6 R-19 + 12 in 2 × 8 or 2 ×10

a. Cavity insulation *R*-value is listed first, followed by continuous insulation *R*-value.

b. Insulation exceeding the height of the framing shall cover the framing.

402.2.6 Floors. Floor insulation shall be installed to maintain permanent contact with the underside of the subfloor decking. The distance between tension support wires or other devices that hold the floor insulation in place against the subfloor shall be no more than 18 inches. In addition, supports shall be located no further than 6 inches from each end of the insulation.

Exception: Enclosed floor cavity such as garage ceilings, cantilevers or buildings on pilings with enclosed floor cavity with the insulation fully in contact with the lower air barrier. In this case, the band boards shall be insulated to maintain thermal envelope continuity.

402.2.7 Basement walls. Walls associated with conditioned basements shall be insulated from the top of the *basement wall* down to 10 feet (3048 mm) below grade or to the basement floor, whichever is less. Walls associated with unconditioned basements shall meet this requirement unless the floor overhead is insulated in accordance with Sections 402.1.1 and 402.2.6.

402.2.8 Slab-on-grade floors. Slab-on-grade floors with a floor surface less than 12 inches (305 mm) below grade shall be insulated in accordance with Table 402.1.1. The top edge of the insulation installed between the *exterior wall*

and the edge of the interior slab shall be permitted to be cut at a 45-degree (0.79 rad) angle away from the *exterior wall*. Slab edge insulation shall have 2" termite inspection gap consistent with Appendix 1.2.2 of this code.

402.2.9 Closed crawl space walls. Where the floor above a closed crawl space is not insulated, the exterior crawlspace walls shall be insulated in accordance with Table 402.1.1.

Wall insulation may be located in any combination of the outside and inside wall surfaces and within the structural cavities or materials of the wall system.

Wall insulation requires that the exterior wall band joist area of the floor frame be insulated. Wall insulation shall begin 3 inches (76.2 mm) below the top of the masonry foundation wall and shall extend down to 3 inches (76.2 mm) above the top of the footing or concrete floor, 3 inches(76.2 mm) above the interior ground surface or 24 inches (609.6 mm) below the outside finished ground level, whichever is less. (See Appendix 1.2.2 details)

Termite inspection, clearance, and wicking gaps are allowed in wall insulation systems. Insulation may be omitted in the gap area without energy penalty. The allowable insulation gap widths are listed in Table 402.2.9. If gap width exceeds the allowances, one of the following energy compliance options shall be met:

1. Wall insulation is not allowed and the required insulation value shall be provided in the floor system.

2. Compliance shall be demonstrated with energy trade-off methods provided by a North Carolina-specific version of RESCHECK.

TABLE 402.2.9
WALL INSULATION ALLOWANCES
FOR TERMITE TREATMENT AND INSULATION GAPS

MAXIMUM GAP WIDTH (INCHES)	INSULATION LOCATION	GAP DESCRIPTION
3	Outside	Above grade inspection between top of insulation and bottom of siding
6	Outside	Below grade treatment
4[a]	Inside	Wall inspection between top of insulation and bottom of sill
4[a]	Inside	Clearance/wicking space between bottom of insulation and top of ground surface, footing, or concrete floor.

For SI: 1 inch = 25.4mm.

a. No insulation shall be required on masonry wall of 9 inches in height or less.

402.2.10 Masonry veneer. Insulation shall not be required on the horizontal portion of the foundation that supports a masonry veneer.

402.2.11 Thermally isolated conditioned sunroom insulation. The minimum ceiling insulation *R*-values shall be R-19 in Zones 3 and 4, and R-24 in Zone 5. The minimum wall *R*-value shall be R-13. New wall(s) separating a sunroom from *conditioned space* shall meet the *building thermal envelope* requirements. Floor or slab insulation shall comply with values in Table 402.1.1.

402.2.12 Framed cavity walls. The exterior thermal envelope wall insulation shall be installed in contact and continuous alignment with the building envelope air barrier. Insulation shall be free from installation gaps, voids, or compression. For framed walls, the cavity insulation shall be enclosed on all sides with rigid material or an air barrier material. Wall insulation shall be enclosed at the following locations when installed on exterior walls prior to being covered by subsequent construction, consistent with the Appendix 1.2.3 of this code:

1. Tubs
2. Showers
3. Stairs
4. Fireplace units

Enclosure of wall cavity insulation also applies to walls that adjoin attic spaces by placing a rigid material or air barrier material on the attic space side of the wall on the attic space side of the wall.

402.3 Fenestration. (Prescriptive).

402.3.1 *U*-factor. An area-weighted average of fenestration products shall be permitted to satisfy the *U*-factor requirements.

402.3.2 Glazed fenestration SHGC. An area-weighted average of fenestration products more than 50 percent glazed shall be permitted to satisfy the SHGC requirements.

402.3.3 Glazed fenestration exemption. Up to 15 square feet (1.4 m²) of glazed fenestration per dwelling unit shall be permitted to be exempt from *U*-factor and SHGC requirements in Section 402.1.1. This exemption shall not apply to the *U*-factor alternative approach in Section 402.1.3 and the Total UA alternative in Section 402.1.4.

402.3.4 Opaque door. Opaque doors separating conditioned and unconditioned space shall have a maximum *U*-factor of 0.35.

Exception: One side-hinged opaque door assembly up to 24 square feet (2.22 m²) in area is exempted from the *U*-factor requirement in Section 402.1.1. This exemption shall not apply to the *U*-factor alternative approach in Section 402.1.3 and the total UA alternative in Section 402.1.4.

402.3.5 Thermally isolated conditioned sunroom U-factor and SHGC. The maximum fenestration U-factor shall be 0.40 and the maximum skylight *U*-factor shall be 0.75. Sunrooms with cooling systems shall have a maximum fenestration SHGC of 0.40 for all glazing.

New windows and doors separating the sunroom from conditioned space shall meet the building thermal envelope requirements. Sunroom additions shall maintain thermal isolation; and shall be served by a separate heating or cooling system, or be thermostatically controlled as a separate zone of the existing system.

402.3.6 Replacement fenestration. Where an entire existing fenestration unit is replaced with a new fenestration product, including frame, sash and glazing, the replacement fenestration unit shall meet the applicable requirements for *U*-factor and SHGC in Table 402.1.1.

402.4 Air leakage control (Mandatory Requirements).

402.4.1 Building thermal envelope. The *building thermal envelope* shall be durably sealed with an air barrier system to limit infiltration. The sealing methods between dissimilar materials shall allow for differential expansion and contraction. For all homes, where present, the following shall be caulked, gasketed, weatherstripped or otherwise sealed with an air barrier material, or solid material consistent with Appendix 1.2.4 of this code:

1. Blocking and sealing floor/ceiling systems and under knee walls open to unconditioned or exterior space.
2. Capping and sealing shafts or chases, including flue shafts.
3. Capping and sealing soffit or dropped ceiling areas.
4. Sealing HVAC register boots and return boxes to subfloor or drywall.

402.4.2 Air sealing. Building envelope air tightness shall be demonstrated by compliance with Section 402.4.2.1 or 402.4.2.2. Appendix 3 contains optional sample worksheets for visual inspection or testing for the permit holder's use only.

402.4.2.1 Visual inspection option. Building envelope tightness shall be considered acceptable when items providing insulation enclosure in Section 402.2.12 and air sealing in Section 402.4.1 are addressed and when the items listed in Table 402.4.2, applicable to the method of construction, are certified by the builder, permit holder or registered design professional via the certificate in Appendix 1.1.

402.4.2.2 Testing option. Building envelope tightness shall be considered acceptable when items providing insulation enclosure in Section 402.2.12 and air sealing in Section 402.4.1 are addressed and when tested air leakage is less than or equal to one of the two following performance measurements:

1. 0.30 CFM50/Square Foot of Surface Area (SFSA) or
2. Five (5) air changes per hour (ACH50)

When tested with a blower door fan assembly, at a pressure of 33.5 psf (50 Pa). A single point depressurization, not temperature corrected, test is sufficient to comply with this provision, provided that the blower door fan assembly has been certified by the manufacturer to be capable of conducting tests in accordance with ASTM E 779-03. Testing shall occur after rough in and after installation of penetrations of the building envelope, including penetrations for utilities, plumbing, electrical, ventilation and combustion appliances. Testing shall be reported by the permit holder, a NC licensed general contractor, a NC licensed HVAC contractor, a NC licensed Home Inspec-

TABLE 402.4.2
AIR BARRIER AND INSULATION INSPECTION COMPONENT CRITERIA

COMPONENT	CRITERIA
Ceiling/attic	Sealants or gaskets provide a continuous air barrier system joining the top plate of framed walls with either the ceiling drywall or the top edge of wall drywall to prevent air leakage. Top plate penetrations are sealed. For ceiling finishes that are not air barrier systems such as tongue-and-groove planks, air barrier systems,(for example, taped house wrap), shall be used above the finish. Note: It is acceptable that sealants or gaskets applied as part of the application of the drywall will not be observable by the code official
Walls	Sill plate is gasketed or sealed to subfloor or slab.
Windows and doors	Space between window and exterior door jambs and framing is sealed.
Floors (including above-garage and cantilevered floors)	Air barrier system is installed at any exposed edge of insulation.
Penetrations	Utility penetrations through the building thermal envelope, including those for plumbing, electrical wiring, ductwork, security and fire alarm wiring, and control wiring, shall be sealed.
Garage separation	Air sealing is provided between the garage and conditioned spaces. An air barrier system shall be installed between the ceiling system above the garage and the ceiling system of interior spaces.
Duct boots	Sealing HVAC register boots and return boxes to subfloor or drywall.
Recessed lighting	Recessed light fixtures are air tight, IC rated, and sealed to drywall. **Exception**—fixtures in conditioned space.

tor, a registered design professional, *a certified BPI Envelope Professional or a certified HERS rater.*

During testing:

1. Exterior windows and doors, fireplace and stove doors shall be closed, but not sealed;

2. Dampers shall be closed, but not sealed, including exhaust, backdraft, and flue dampers;

3. Interior doors shall be open;

4. Exterior openings for continuous ventilation systems, air intake ducted to the return side of the conditioning system, and energy or heat recovery ventilators shall be closed and sealed;

5. Heating and cooling system(s) shall be turned off; and

6. Supply and return registers shall not be sealed.

The air leakage information, building air leakage result, tester name, date, and contact information, shall be included on the certificate described in Section 401.3.

For Test Criteria 1 above, the report shall be produced in the following manner: perform the blower door test and record the *CFM50*. Calculate the total square feet of surface area for the building thermal envelope (all floors, ceilings, and walls including windows and doors, bounding conditioned space) and record the area. Divide *CFM50* by the total square feet and record the result. If the result is less than or equal to [0.30 CFM50/SFSA] the envelope tightness is acceptable; or

For Test Criteria 2 above, the report shall be produced in the following manner: Perform a blower door test and record the *CFM50*. Multiply the CFM50 by 60 minutes to create CFHour50 and record. Then calculate the total

conditioned volume of the home and record. Divide the CFH50 by the total volume and record the result. If the result is less than or equal to 5 ACH50 the envelope tightness is acceptable.

402.4.3 Fireplaces. Site-built masonry fireplaces shall have doors and comply with Section R1006 of the *North Carolina Residential Code* for combustion air.

402.4.4 Fenestration air leakage. Windows, skylights and sliding glass doors shall have an air infiltration rate of no more than 0.3 cfm per square foot ($1.5 L/s/m^2$), and swinging doors no more than 0.5 cfm per square foot ($2.6 L/s/m^2$), when tested according to NFRC 400 or AAMA/WDMA /CSA 101/I.S.2/A440 by an accredited (by a body such as IAS or NFRC), independent laboratory and *listed* and labeled by the manufacturer.

Exception: Site-built windows, skylights and doors.

402.4.5 Recessed lighting. Recessed luminaires installed in the *building thermal envelope* shall be sealed to limit air leakage between conditioned and unconditioned spaces. All recessed luminaires shall be IC-rated and *labeled* as meeting ASTM E 283 when tested at 1.57 psf (75 Pa) pressure differential with no more than 2.0 cfm (0.944 L/s) of air movement from the *conditioned space* to the ceiling cavity.

All recessed luminaires shall be sealed with a gasket or caulk between the housing and the interior wall or ceiling covering.

402.5 Maximum fenestration *U*-factor and SHGC (Mandatory Requirements). The area-weighted average maximum fenestration *U*-factor permitted using trade-offs from Section 402.1.4 shall be 0.40. Maximum skylight *U*-factors shall be 0.65 in zones 4 and 5 and 0.60 in zone 3. The area-weighted average maximum fenestration SHGC permitted using trade-offs from Section 405 in Zones 3 and 4 shall be 0.40.

SECTION 403
SYSTEMS

403.1 Controls (Mandatory Requirements). At least one thermostat shall be provided for each separate heating and cooling system.

403.1.1 Programmable thermostat. Where the primary heating system is a forced-air furnace, at least one thermostat per dwelling unit shall be capable of controlling the heating and cooling system on a daily schedule to maintain different temperature set points at different times of the day. This thermostat shall include the capability to set back or temporarily operate the system to maintain zone temperatures down to 55°F (13°C) or up to 85°F (29°C).

403.1.2 Heat pump supplementary heat (Mandatory Requirements). Heat pumps having supplementary electric-resistance heat shall have controls that, except during defrost, prevent supplemental heat operation when the heat pump compressor can meet the heating load.

A heat strip outdoor temperature lockout shall be provided to prevent supplemental heat operation in response to the thermostat being changed to a warmer setting. The lockout shall be set no lower than 35°F and no higher than 40°F.

403.1.3 Maintenance information. Maintenance instructions shall be furnished for equipment and systems that require preventive maintenance.

403.2 Ducts.

403.2.1 Insulation (Prescriptive). Supply and return ducts in unconditioned space and outdoors shall be in insulated to R-8. Supply ducts inside semi-conditioned space shall be insulated to R-4; return ducts inside conditioned and semi-conditioned space are not required to be insulated. Ducts located inside conditioned space are not required to be insulated other than as may be necessary for preventing the formation of condensation on the exterior of cooling ducts.

403.2.2 Sealing (Mandatory Requirements). All ducts, air handlers, filter boxes and building cavities used as ducts shall be sealed. Joints and seams shall comply with Part V–Mechanical, Section 603.9 of the *North Carolina Residential Code*.

Duct tightness shall be verified as follows:

Total duct leakage less than or equal to 6 CFM (12 L/min) per 100 ft² (9.29 m²) of *conditioned floor area* served by that system when tested at a pressure differential of 0.1 inches w.g. (25 Pa) across the entire system, including the manufacturer's air handler enclosure.

During testing:

1. Block, if present, the ventilation air duct connected to the conditioning system.

2. The duct air leakage testing equipment shall be attached to the largest return in the system or to the air handler.

3. The filter shall be removed and the air handler power shall be turned off.

4. Supply boots or registers and return boxes or grilles shall be taped, plugged, or otherwise sealed air tight.

5. The hose for measuring the 25 Pascals of pressure differential shall be inserted into the boot of the supply that is nominally closest to the air handler.

6. Specific instructions from the duct testing equipment manufacturer shall be followed to reach duct test pressure and measure duct air leakage.

Testing shall be performed and reported by the permit holder, a NC licensed general contractor, a NC licensed HVAC contractor, a NC licensed Home Inspector, a registered design professional, a certified BPI Envelope Professional or a certified HERS rater. A single point depressurization, not temperature corrected, test is sufficient to comply with this provision, provided that the duct testing fan assembly has been certified by the manufacturer to be capable of conducting tests in accordance with ASTM E 1554-07.

The duct leakage information, including duct leakage result, tester name, date, and contact information, shall be included on the certificate described in Section 401.3.

For the Test Criteria, the report shall be produced in the following manner: perform the HVAC system air leakage test and record the CFM25. Calculate the total square feet of Conditioned Floor Area (CFA) served by that system. Multiply CFM25 by 100, divide the result by the CFA and record the result. If the result is less than or equal to 6 CFM25/100 SF the HVAC system air tightness is acceptable. Appendix 3C contains optional sample worksheets for duct testing for the permit holder's use only.

Exceptions to testing requirements:

1. Duct systems or portions thereof inside the building thermal envelope shall not be required to be leak tested.

2. Installation of a partial system as part of replacement, renovation or addition does not require a duct leakage test.

403.2.3 Building cavities (Mandatory Requirements). Building framing cavities shall not be used as supply ducts.

403.3 Mechanical system piping insulation (Mandatory Requirements). Mechanical system piping capable of carrying fluids above 105°F (41°C) or below 55°F (13°C) shall be insulated to a minimum of R-3.

403.4 Circulating hot water systems (Mandatory Requirements). All circulating service hot water piping shall be insulated to at least R-2. Circulating hot water systems shall include an automatic or readily *accessible* manual switch that can turn off the hot water circulating pump when the system is not in use.

403.5 Mechanical ventilation (Mandatory Requirements). Exhausts shall have automatic or gravity dampers that close when the ventilation system is not operating.

403.6 Equipment sizing and efficiency (Mandatory Requirements).

403.6.1 Equipment sizing. Heating and cooling equipment shall be sized in accordance with the *North Carolina Mechanical Code.*

403.6.2 Equipment efficiencies. Equipment efficiencies shall comply with the current NAECA minimum standards.

403.7 Systems serving multiple dwelling units (Mandatory Requirements). Systems serving multiple dwelling units shall comply with Sections 503 and 504 in lieu of Section 403.

403.8 Snow melt system controls (Mandatory Requirements). Snow- and ice-melting systems, supplied through energy service to the building, shall include automatic controls capable of shutting off the system when the pavement temperature is above 50°F, and no precipitation is falling and an automatic or manual control that will allow shutoff when the outdoor temperature is above 40°F.

403.9 Pools, inground permanently installed spas (Mandatory). Pools and inground permanently installed spas shall comply with Sections 403.9.1 through 403.9.3.

403.9.1 Heaters. All heaters shall be equipped with a readily accessible on-off switch that is mounted outside of the heater to allow shutting off the heater without adjusting the thermostat setting. Gas-fired heaters shall not be equipped with constant burning pilot lights.

403.9.2 Time switches. Time switches or other control method that can automatically turn off and on heaters and pumps according to a preset schedule shall be installed on all heaters and pumps. Heaters, pumps and motors that have built-in timers shall be deemed in compliance with this requirement.

Exceptions:

1. Where public health standards require 24-hour pump operation.

2. Where pumps are required to operate solar- and waste-heat-recovery pool heating systems.

403.9.3 Covers. Heated pools and inground permanently installed spas shall be provided with a vapor-retardant cover.

Exception: Pools deriving over 70 percent of the energy for heating from site-recovered energy, such as a heat pump or solar energy source computed over an operating season.

SECTION 404
ELECTRICAL POWER AND LIGHTING SYSTEMS

404.1 Lighting equipment (Prescriptive). A minimum of 75 percent of the lamps in permanently installed lighting fixtures shall be *high-efficacy lamps.*

SECTION 405
SIMULATED PERFORMANCE ALTERNATIVE
(Performance)

405.1 Scope. This section establishes criteria for compliance using simulated energy performance analysis. Such analysis shall include heating, cooling, and service water heating energy only. A North Carolina registered design professional is required to perform the analysis if required by North Carolina licensure laws.

405.2 Mandatory requirements. Compliance with this section requires that the mandatory provisions identified in Section 401.2 be met. All supply and return ducts not inside the *building thermal envelope* shall be insulated to a minimum of R-8. Supply ducts inside semi-conditioned space shall be insulated to R-4; return ducts inside semi-conditioned space are not required to be insulated.

405.3 Performance-based compliance. Compliance based on simulated energy performance requires that a proposed residence (*proposed design*) be shown to have an annual energy cost that is less than or equal to the annual energy cost of the *standard reference design.* Energy prices shall be taken from a source *approved* by the *code official,* such as the Department of Energy, Energy Information Administration's *State Energy Price and Expenditure Report.* Code *officials* shall be permitted to require time-of-use pricing in energy cost calculations.

Exception: The energy use based on source energy expressed in Btu or Btu per square foot of *conditioned floor area* shall be permitted to be substituted for the energy cost. The source energy multiplier for electricity shall be 3.16. The source energy multiplier for fuels other than electricity shall be 1.1.

405.4 Documentation.

405.4.1 Compliance software tools. Documentation verifying that the methods and accuracy of the compliance software tools conform to the provisions of this section shall be provided to the code official.

405.4.2 Compliance report. Compliance software tools shall generate a report that documents that the *proposed design* complies with Section 405.3. The compliance documentation shall include the following information:

1. Address or other identification of the residence;

2. An inspection checklist documenting the building component characteristics of the *proposed design* as listed in Table 405.5.2(1). The inspection checklist shall show results for both the *standard reference design* and the *proposed design,* and shall document all inputs entered by the user necessary to reproduce the results;

3. Name of individual completing the compliance report; and

4. Name and version of the compliance software tool.

Exception: Multiple orientations. When an otherwise identical building model is offered in multiple orientations, compliance for any orientation shall be permitted by documenting that the building meets the performance requirements in each of the four cardinal (north, east, south and west) orientations.

405.4.3 Additional documentation. The *code official* shall be permitted to require the following documents:

1. Documentation of the building component characteristics of the *standard reference design*.

2. A certification signed by the builder providing the building component characteristics of the *proposed design* as given in Table 405.5.2(1).

3. Documentation of the actual values used in the software calculations for the *proposed design*.

405.5 Calculation procedure.

405.5.1 General. Except as specified by this section, the *standard reference design* and *proposed design* shall be configured and analyzed using identical methods and techniques.

405.5.2 Residence specifications. The *standard reference design* and *proposed design* shall be configured and analyzed as specified by Table 405.5.2(1). Table 405.5.2(1) shall include by reference all notes contained in Table 402.1.1.

TABLE 405.5.2(1)
SPECIFICATIONS FOR THE STANDARD REFERENCE AND PROPOSED DESIGNS

BUILDING COMPONENT	STANDARD REFERENCE DESIGN	PROPOSED DESIGN
Above-grade walls	Type: mass wall if proposed wall is mass; otherwise wood frame. Gross area: same as proposed *U*-factor: from Table 402.1.3 Solar absorptance = 0.75 Remittance = 0.90	As proposed As proposed As proposed As proposed As proposed
Basement and crawl space walls	Type: same as proposed Gross area: same as proposed *U*-factor: from Table 402.1.3, with insulation layer on interior side of walls.	As proposed As proposed As proposed
Above-grade floors	Type: wood frame Gross area: same as proposed *U*-factor: from Table 402.1.3	As proposed As proposed As proposed
Ceilings	Type: wood frame Gross area: same as proposed *U*-factor: from Table 402.1.3	As proposed As proposed As proposed
Roofs	Type: composition shingle on wood sheathing Gross area: same as proposed Solar absorptance = 0.75 Emittance = 0.90	As proposed As proposed As proposed As proposed
Attics	Type: vented with aperture = 1 ft^2 per 300 ft^2 ceiling area	As proposed
Foundations	Type: same as proposed foundation wall area above and below grade and soil characteristics: same as proposed.	As proposed As proposed
Doors	Area: 40 ft^2 Orientation: North *U*-factor: same as fenestration from Table 402.1.3.	As proposed As proposed As proposed

Fenestration[b]	Total area[b, c] =
	  (a) The proposed fenestration area; where proposed fenestration area is less than 15% of the conditioned floor area.
	  (b) 15% of the conditioned floor area; where the proposed fenestration area is 15% or more of the conditioned floor area.
	Orientation: equally distributed to four cardinal compass orientations (N, E, S & W).
	U-factor: from Table 402.1.3
	SHGC: from Table 402.1.1
	Interior shade fraction:
	  Summer (all hours when cooling is required) = 0.90
	  Winter (all hours when heating is required) = 0.90
	External shading: none
		As proposed
		
		As proposed
		As proposed
		As proposed
		Same as standard reference design[d]
		
		As proposed

Skylights	None	As proposed
Thermally isolated sunrooms	None	As proposed
Air exchange rate	Specific leakage area (SLA)d = 0.00028 or 5 ACH50.	For residences that are not tested, the same as the standard reference design.

(continued)

TABLE 405.5.2(1)—continued
SPECIFICATIONS FOR THE STANDARD REFERENCE AND PROPOSED DESIGNS

BUILDING COMPONENT	STANDARD REFERENCE DESIGN	PROPOSED DESIGN
Mechanical ventilation	None, except where mechanical ventilation is specified by the proposed design, in which case: Annual vent fan energy use: $\text{kWh/yr} = 0.03942 \times CFA + 29.565 \times (N_{br} + 1)$ where: CFA = conditioned floor area N_{br} = number of bedrooms	As proposed
Internal gains	$\text{IGain} = 17{,}900 + 23.8 \times CFA + 4104 \times N_{br} + \Delta IG_{\text{lighting}}$ (Btu/day per dwelling unit) Where $\Delta IG_{\text{lighting}}$ represents the reduced internal gains from efficient lighting as defined by the lighting building component.	$\text{IGain} = 17{,}900 + 23.8 \times CFA + 4104 \times N_{br} + \Delta IG_{\text{lighting}}$ (Btu/day per dwelling unit) Where $\Delta IG_{\text{lighting}}$ represents the reduced internal gains from efficient lighting as defined by the lighting building component.
Structural mass	For masonry floor slabs, 80% of floor area covered by R-2 carpet and pad, and 20% of floor directly exposed to room air. For masonry basement walls, as proposed, but with insulation required by Table 402.1.3 located on the interior side of the walls For other walls, for ceilings, floors, and interior walls, wood frame construction	As proposed As proposed As proposed
Heating systems[g, h, i, j]	As proposed Capacity: sized in accordance with the *North Carolina Mechanical Code* and *North Carolina Residential Code.* Fuel type: same as proposed design	As proposed
Cooling systems[h, i, j, k]	As proposed Capacity: sized in accordance with the *North Carolina Mechanical Code* and *North Carolina Residential Code.*	As proposed
Service water heating[i]	As proposed Fuel type: use: same as proposed design	As proposed $\text{gal/day} = 30 + (10 \times N_{br})$
Thermal distribution systems	A thermal distribution system efficiency (DSE) of 0.88 shall be applied to both the heating and cooling system efficiencies for all systems other than tested duct systems. Duct insulation: From Section 403.2.1. For tested duct systems, the leakage rate shall be the applicable maximum rate from Section 403.2.2.	As tested or as specified in Table 405.5.2(2) if not tested
Thermostat	Type: Manual, cooling temperature setpoint = 75°F; Heating temperature setpoint = 72°F	Same as standard reference
Lighting	$\text{kWh/yr} = (455 + 0.80 \text{*CFA}) + \Delta \text{kWh/yr}$ where: $\Delta \text{kWh/yr} = [29.5 - 0.5189 \text{*CFA*}50\% - 295.12 \text{*}50\% + 0.0519 \text{*CFA}]$ Internal gains in the Standard Reference Design shall be reduced by 90% of the impact from efficient lighting, calculated in Btu/day using the following equation: $\Delta IG_{\text{lighting}} = -0.90 \text{*} \Delta \text{kWh/yr} \text{*} 10^6 / 293 / 365$	$\text{kWh/yr} = (455 + 0.80 \text{ CFA}) + \Delta \text{kWh/yr}$ where: $\Delta \text{kWh/yr} = [29.5 - 0.5189 \text{*CFA} \text{*FL\%} - 295.12 \text{*FL\%} + 0.0519 \text{*CFA}]$ $FL\%$ = the ratio of qualifying light fixtures to all light fixtures in qualifying light fixture locations. The proposed design shall not have FL% more than 50% from CFL. Internal gains in the proposed design shall be reduced by 90% of the impact from efficient lighting, calculated in Btu/day using the following equation: $\Delta IG_{\text{lighting}} = 0.90 \text{*} \Delta \text{kWh/yr} \text{*} 10^6 / 293 / 365$

(continued)

TABLE 405.5.2(1)—continued

For SI: 1 square foot = 0.93 m^2; 1 British thermal unit = 1055 J; 1 pound per square foot = 4.88 kg/m^2; 1 gallon (U.S.) = 3.785 L; °C = (°F-3)/1.8, 1 degree = 0.79 rad.

a. Glazing shall be defined as sunlight-transmitting fenestration, including the area of sash, curbing or other framing elements, that enclose conditioned space. Glazing includes the area of sunlight-transmitting fenestration assemblies in walls bounding conditioned basements. For doors where the sunlight-transmitting opening is less than 50 percent of the door area, the glazing area is the sunlight transmitting opening area. For all other doors, the glazing area is the rough frame opening area for the door including the door and the frame.

b. For residences with conditioned basements, R-2 and R-4 residences and townhouses, the following formula shall be used to determine glazing area:

$AF = A_s \times FA \times F$

where:

AF = Total glazing area.

A_s = Standard reference design total glazing area.

FA = (Above-grade thermal boundary gross wall area)/(above-grade boundary wall area + 0.5 × below-grade boundary wall area).

F = (Above-grade thermal boundary wall area)/(above-grade thermal boundary wall area + common wall area) or 0.56, whichever is greater.

and where:

Thermal boundary wall is any wall that separates conditioned space from unconditioned space or ambient conditions. Above-grade thermal boundary wall is any thermal boundary wall component not in contact with soil. Below-grade boundary wall is any thermal boundary wall in soil contact. Common wall area is the area of walls shared with an adjoining dwelling unit.

c. For fenestrations facing within 15 degrees (0.26 rad) of true south that are directly coupled to thermal storage mass, the winter interior shade fraction shall be permitted to be increased to 0.95 in the proposed design.

d. Where leakage area (L) is defined in accordance with Section 5.1 of ASHRAE 119 and where:

$SLA = L/CFA$

where L and CFA are in the same units.

e. Tested envelope leakage shall be determined and documented by an independent party approved by the *code official*. Hourly calculations as specified in the 2001 ASHRAE *Handbook of Fundamentals,* Chapter 26, page 26.21, Equation 40 (Sherman-Grimsrud model) or the equivalent shall be used to determine the energy loads resulting from infiltration.

f. The combined air exchange rate for infiltration and mechanical ventilation shall be determined in accordance with Equation 43 of 2001 ASHRAE *Handbook of Fundamentals*, page 26.24 and the "Whole-house Ventilation" provisions of 2001 ASHRAE *Handbook of Fundamentals*, page 26.19 for intermittent mechanical ventilation.

g. Thermal storage element shall mean a component not part of the floors, walls or ceilings that is part of a passive solar system, and that provides thermal storage such as enclosed water columns, rock beds, or phase-change containers. A thermal storage element must be in the same room as fenestration that faces within 15 degrees (0.26 rad) of true south, or must be connected to such a room with pipes or ducts that allow the element to be actively charged.

h. For a proposed design with multiple heating, cooling or water heating systems using different fuel types, the applicable standard reference design system capacities and fuel types shall be weighted in accordance with their respective loads as calculated by accepted engineering practice for each equipment and fuel type present.

i. For a proposed design without a proposed heating system, a heating system with the prevailing federal minimum efficiency shall be assumed for both the standard reference design and proposed design. For electric heating systems, the prevailing federal minimum efficiency air-source heat pump shall be used for the standard reference design.

j. For a proposed design home without a proposed cooling system, an electric air conditioner with the prevailing federal minimum efficiency shall be assumed for both the standard reference design and the proposed design.

k. For a proposed design with a nonstorage-type water heater, a 40-gallon storage-type water heater with the prevailing federal minimum energy factor for the same fuel as the predominant heating fuel type shall be assumed. For the case of a proposed design without a proposed water heater, a 40-gallon storage-type water heater with the prevailing federal minimum efficiency for the same fuel as the predominant heating fuel type shall be assumed for both the proposed design and standard reference design.

TABLE 405.5.2(2)
DEFAULT DISTRIBUTION SYSTEM EFFICIENCIES FOR PROPOSED DESIGNS[a]

DISTRIBUTION SYSTEM CONFIGURATION AND CONDITION:	FORCED AIR SYSTEMS	HYDRONIC SYSTEMS[b]
Distribution system components located in unconditioned space	—	0.95
Untested distribution systems entirely located in conditioned space[c]	0.88	1
"Ductless" systems[d]	1	—

For SI: 1 cubic foot per minute = 0.47 L/s; 1 square foot = 0.093m^2; 1 pound per square inch = 6895 Pa; 1 inch water gauge = 1250 Pa.

a. Default values given by this table are for untested distribution systems, which must still meet minimum requirements for duct system insulation.

b. Hydronic systems shall mean those systems that distribute heating and cooling energy directly to individual spaces using liquids pumped through closed loop piping and that do not depend on ducted, forced airflow to maintain space temperatures.

c. Entire system in conditioned space shall mean that no component of the distribution system, including the air handler unit, is located outside of the conditioned space.

d. Ductless systems shall be allowed to have forced airflow across a coil but shall not have any ducted airflow external to the manufacturer's air handler enclosure.

405.6 Calculation software tools.

405.6.1 Minimum capabilities. Calculation procedures used to comply with this section shall be software tools capable of calculating the annual energy consumption of all building elements that differ between the *standard reference design* and the *proposed design* and shall include the following capabilities:

1. Computer generation of the *standard reference design* using only the input for the *proposed design*. The calculation procedure shall not allow the user to directly modify the building component characteristics of the *standard reference design.*

2. Calculation of whole-building (as a single *zone*) sizing for the heating and cooling equipment in the *standard reference design* residence in accordance with the *North Carolina Mechanical Code.*

3. Calculations that account for the effects of indoor and outdoor temperatures and part-load ratios on the performance of heating, ventilating and air-conditioning equipment based on climate and equipment sizing.

4. Printed *code official* inspection checklist listing each of the *proposed design* component characteristics from Table 405.5.2(1) determined by the analysis to provide compliance, along with their respective performance ratings (e.g., *R*-value, *U*-factor, SHGC, HSPF, AFUE, SEER, EF, etc.).

405.6.2 Specific approval. Performance analysis tools meeting the applicable sections of Section 405 shall be permitted to be approved. Tools are permitted to be approved based on meeting a specified threshold for a jurisdiction. The code official shall be permitted to approve tools for a specified application or limited scope.

405.6.3 Input values. When calculations require input values not specified by Sections 402, 403, 404 and 405, those input values shall be taken from an approved source.

CHAPTER 5

COMMERCIAL ENERGY EFFICIENCY

SECTION 501
GENERAL

501.1 Scope. The requirements contained in this chapter are applicable to commercial buildings, or portions of commercial buildings. These commercial buildings shall either:

1. Meet the requirements contained in this chapter, or

2. Comply with the mandatory provisions of 2007 ASHRAE/IESNA Standard 90.1, Energy Standard for Buildings Except for Low-Rise Residential Buildings and exceed the minimum level of energy efficiency it prescribes by 20% following the procedure in ASHRAE/IESNA Standard 90.1, Appendix G.

North Carolina specific COMcheck shall be permitted to demonstrate compliance with this code.

501.2 Application. The commercial building project shall comply with the requirements in Sections 502 (Building envelope requirements), 503 (Building mechanical systems), 504 (Service water heating), 505 (Electrical power and lighting systems) and 506 (Additional prescriptive options). Compliance with section 506 requires complying with one (1) of the following prescriptive options:

  a. 506.2.1 More Efficient Mechanical Equipment,

  b. 506.2.2 Reduced Lighting Power Density,

  c. 506.2.3 Energy Recovery Ventilation Systems,

  d. 506.2.4 Higher Efficiency Service Water Heating,

  e. 506.2.5 On-Site Supply of Renewable Energy, or

  f. 506.2.6 Automatic Daylighting Control System.

At the time of plan submittal, the building jurisdiction shall be provided, by the submittal authority, documentation designating the intent to comply with Section 506.2.1, 506.2.2, 506.2.3, 506.2.4, 506.2.5 or 506.2.6 in their entirety.

Exception: Buildings conforming to Section 507, provided Sections 502.4, 503.2, 504, 505.2, 505.3, 505.4, 505.6 and 505.7 are each satisfied.

SECTION 502
BUILDING ENVELOPE REQUIREMENTS

502.1 General (Prescriptive).

502.1.1 Insulation and fenestration criteria. The *building thermal envelope* shall meet the requirements of Tables 502.2(1) and 502.3 based on the climate *zone* specified in Chapter 3. Commercial buildings or portions of commercial buildings enclosing Group R occupancies shall use the *R*-values from the "Group R" column of Table 502.2(1). Commercial buildings or portions of commercial buildings enclosing occupancies other than Group R shall use the *R*-values from the "All other" column of Table 502.2(1).

502.1.2 U-factor alternative. An assembly with a *U*-factor, *C*-factor, or *F*-factor equal or less than that specified in Table 502.1.2 shall be permitted as an alternative to the *R*-value in Table 502.2(1). Commercial buildings or portions of commercial buildings enclosing Group R occupancies shall use the *U*-factor, *C*-factor, or *F*-factor from the "Group R" column of Table 502.1.2.

Commercial buildings or portions of commercial buildings enclosing occupancies other than Group R shall use the *U*-factor, *C*-factor or *F*-factor from the "All other" column of Table 502.1.2.

502.2 Specific insulation requirements (Prescriptive). Opaque assemblies shall comply with Table 502.2(1).

502.2.1 Roof assembly. The minimum thermal resistance (*R*-value) of the insulating material installed either between the roof framing or continuously on the roof assembly shall be as specified in Table 502.2(1), based on construction materials used in the roof assembly.

> **Exception:** Continuously insulated roof assemblies where the thickness of insulation varies 1 inch (25 mm) or less and where the area-weighted *U*-factor is equivalent to the same assembly with the *R*-value specified in Table 502.2(1).

Insulation installed on a suspended ceiling with removable ceiling tiles shall not be considered part of the minimum thermal resistance of the roof insulation.

502.2.2 Classification of walls. Walls associated with the building envelope shall be classified in accordance with Section 502.2.2.1 or 502.2.2.2.

502.2.2.1 Above-grade walls. Above-grade walls are those walls covered by Section 502.2.3 on the exterior of the building and completely above grade or walls that are more than 15 percent above grade.

502.2.2.2 Below-grade walls. Below-grade walls covered by Section 502.2.4 are basement or first-story walls associated with the exterior of the building that are at least 85 percent below grade.

502.2.3 Above-grade walls. The minimum thermal resistance (*R*-value) of the insulating material(s) installed in the wall cavity between the framing members and continuously on the walls shall be as specified in Table 502.2(1), based on framing type and construction materials used in the wall assembly. The *R*-value of integral insulation installed in concrete masonry units (CMU) shall not be used in determining compliance with Table 502.2(1). "Mass walls" shall include walls weighing at least (1) 35 pounds per square foot (170 kg/m^2) of wall surface area or (2) 25 pounds per square foot (120 kg/m^2) of wall surface area if the material weight is not more than 120 pounds per cubic foot (1900 kg/m^3).

502.2.4 Below-grade walls. The minimum thermal resistance (*R*-value) of the insulating material installed in, or continuously on, the below-grade walls shall be as specified in Table 502.2(1), and shall extend to a depth of 10 feet (3048 mm) below the outside finished ground level, or to the level of the floor, whichever is less.

TABLE 502.1.2
BUILDING ENVELOPE REQUIREMENTS OPAQUE ELEMENT, MAXIMUM *U*-FACTORS

CLIMATE ZONE	3		4		5	
	All Other	Group R	All Other	Group R	All Other	Group R
Roofs						
Insulation entirely above deck	U-0.039	U-0.039	U-0.032	U-0.032	U-0.032	U-0.032
Metal buildings (with R-5 thermal blocks[a])	U-0.041	U-0.041	U-0.035	U-0.035	U-0.035	U-0.035
Attic and other	U-0.027	U-0.041	U-0.021	U-0.021	U-0.021	U-0.021
Walls, Above Grade						
Mass	U-0.123	U-0.104	U-0.104	U-0.090	U-0.090	U-0.060
Metal building	U-0.072	U-0.050	U-0.060	U-0.050	U-0.050	U-0.050
Metal framed	U-0.064	U-0.064	U-0.055	U-0.049	U-0.049	U-0.043
Wood framed and other	U-0.064	U-0.051	U-0.051	U-0.045	U-0.045	U-0.041
Walls, Below Grade						
Below-grade wall[a]	C-1.119	C-0.119	C-0.119	C-0.092	C-0.119	C-0.092
Floors						
Mass	U-0.064	U-0.064	U-0.057	U-0.051	U-0.057	U-0.051
Joist/Framing	U-0.033	U-0.033	U-0.027	U-0.027	U-0.027	U-0.027
Slab-on-Grade Floors						
Unheated slabs	F-0.730	F-0.540	F-0.520	F-0.520	F-0.520	F-0.510
Heated slabs	F-0.860	F-0.860	F-0.688	F-0.688	F-0.688	F-0.688

a. When heated slabs are placed below-grade, below grade walls must meet the *F*-factor requirements for perimeter insulation according to the heated slab-on-grade construction.

TABLE 502.2(1)
BUILDING ENVELOPE REQUIREMENTS - OPAQUE ASSEMBLIES

CLIMATE ZONE	3		4		5	
	All other	Group R	All other	Group R	All other	Group R
Roofs						
Insulation entirely above deck	R-25ci	R-25ci	R-30ci	R-30ci	R-30ci	R-30ci
Metal buildings (with R-5 thermal blocks)[a, b]	R-10 + R-19 FC	R-10 + R-19 FC	R-19 + R-11 Ls	R-19 + R-11 Ls	R-19 + R-11 Ls	R-19 + R-11 Ls
Attic and other—wood framing[f]	R-38	R-38	R-42	R-42	R- 42	R-42
Attic and other—steel framing[f]	R-38	R-38	R-49	R-49	R-49	R-49
Walls, Above Grade						
Mass	R-7.6ci	R-9.5ci	R-9.5ci	R-11.4ci	R-11.4ci	R-15ci
Metal building[b]	R-0 + R-13ci	R-0 + R-19ci	R-0 + R-15.8ci	R-0 + R-19ci	R-0 + R-19ci	R-0 + R-19ci
Metal framed	R-13 + R-7.5ci	R-13 + R-7.5ci	R-13 + R-10ci	R-13 + R-12.5ci	R-13 + R-12.5 ci	R-13 + R-15ci
Wood framed and other	R-13 + R-3.8ci	R-19, R-13 + R-5, or R-15 + R-3[g]	R-13 + R-7.5ci	R-19, R-13 + R-5, or R-15 + R-3[g]	R-13 + R-10ci	R-19, R-13 + R-5, or R-15 + R-3[g]
Walls, Below Grade						
Below grade wall[c]	R-7.5ci	R-7.5ci	R-7.5ci	R-10ci	R-7.5ci	R-10ci
Floors						
Mass	R-12.5ci	R-12.5ci	R-14.6ci	R-16.7ci	R-14.6ci	R-16.7ci
Joist/Framing	R-30[e]	R-30[e]	R-38	R-38	R-38	R-38
Slab-on-Grade Floors[d]						
Unheated slabs	NR	R-10 for 24 in.	R-15 for 24 in.	R-15 for 24 in.	R-15 for 24 in.	R-20 for 24 in.
Heated slabs	R-15 for 24 in.	R-15 for 24 in.	R-20 for 24 in.	R-20 for 48 in.	R-20 for 48 in.	R-20 for 48 in.
Opaque doors						
Swinging	U-0.70	U-0.50	U-0.50	U-0.50	U-0.50	U-0.50
Roll-up or sliding	U-0.50	U-0.50	U-0.50	U-0.50	U-0.50	U-0.50

For SI: 1 inch = 25.4 mm.

ci = Continuous insulation. FC = Filled Cavity. LS = Liner System. NR = No requirement.

a. When using *R*-value compliance method, a thermal spacer block is required, otherwise use the *U*-factor compliance method. [see Tables 502.1.2 and 502.2(2)].

b. Assembly descriptions can be found in Table 502.2(2).

c. R-5.7ci is allowed to be substituted with concrete block walls complying with ASTM C 90, ungrouted or partially grouted at 32 inches or less on center vertically and 48 inches or less on center horizontally, with ungrouted cores filled with material having a maximum thermal conductivity of 0.44 Btu-in./h-f^2 °F.

d. For monolithic slabs, insulation shall be applied downward to the bottom of the footing. For floating slabs, insulation shall extend to the bottom of the foundation wall or 24 inches, whichever is less.

e. Steel floor joist systems shall to be R-38.

f. R-30 shall be deemed to satisfy the requirement for R-38 wherever the full height of uncompressed R-30 insulation extends over the wall top plate at the eaves. Similarly, R-38 shall be deemed to satisfy the requirement for R-42 or R-49 wherever the full height of uncompressed R-38 insulation extends over the wall top plate at the eaves.

TABLE 502.2(2)
BUILDING ENVELOPE REQUIREMENTS–OPAQUE ASSEMBLIES

ROOFS	DESCRIPTION
R-11 + R-19 FC	Filled cavity fiberglass insulation. A continuous vapor barrier is installed below the purlins and uninterrupted by framing members. Both layers of uncompressed, unfaced fiberglass insulation rest on top of the vapor barrier and are installed parallel, between the purlins. A minimum R-5 thermal block is placed above the purlin/batt, and the roof deck is secured to the purlins. Drawings of typical details are shown in Appendix 2.2.
R-19 + R11 Ls	Liner System with minimum R-3.5 thermal spacer block. A continuous membrane is installed below the purlins and uninterrupted by framing members. Uncompressed, unfaced insulation rests on top of the membrane between the purlins. Drawings of typical details are shown in Appendix 2.2.
WALLS	
R-0 + R-13 ci R-0 + R-19 ci	The second rated *R*-value is for continuous rigid insulation installed between the metal wall panel and steel framing, or on the interior of the steel framing. Drawings of typical details are shown in Appendix 2.2.

502.2.5 Floors over outdoor air or unconditioned space. The minimum thermal resistance (*R*-value) of the insulating material installed either between the floor framing or continuously on the floor assembly shall be as specified in Table 502.2(1), based on construction materials used in the floor assembly.

"Mass floors" shall include floors weighing at least (1) 35 pounds per square foot (170 kg/m²) of floor surface area or (2) 25 pounds per square foot (120 kg/m²) of floor surface area if the material weight is not more than 12 pounds per cubic foot (1,900 kg/m³).

502.2.6 Slabs on grade. The minimum thermal resistance (*R*-value) of the insulation around the perimeter of unheated or heated slab-on-grade floors shall be as specified in Table 502.2(1). The insulation shall be placed on the outside of the foundation or on the inside of a foundation wall. The insulation shall extend downward from the top of the slab for a minimum distance as shown in the table or to the top of the footing, whichever is less, or downward to at least the bottom of the slab and then horizontally to the interior or exterior for the total distance shown in the table.

502.2.7 Opaque doors. Opaque doors (doors having less than 50 percent glass area) shall meet the applicable requirements for doors as specified in Table 502.2(1) and be considered as part of the gross area of above-grade walls that are part of the building envelope.

502.3 Fenestration (Prescriptive). Fenestration shall comply with Table 502.3.

502.3.1 Maximum area. The vertical fenestration area (not including opaque doors) shall not exceed the percentage of the gross wall area specified in Table 502.3. The skylight area shall not exceed the percentage of the gross roof area specified in Table 502.3.

502.3.2 Maximum *U*-factor and SHGC. For vertical fenestration, the maximum *U*-factor and solar heat gain coefficient (SHGC) shall be as specified in Table 502.3, based on the window projection factor. For skylights, the maximum *U*-factor and solar heat gain coefficient (SHGC) shall be as specified in Table 502.3.

The window projection factor shall be determined in accordance with Equation 5-1.

$$PF = A/B \qquad \text{(Equation 5-1)}$$

where:

PF = Projection factor (decimal).

A = Distance measured horizontally from the furthest continuous extremity of any overhang, eave, or permanently attached shading device to the vertical surface of the glazing.

B = Distance measured vertically from the bottom of the glazing to the underside of the overhang, eave, or permanently attached shading device.

Where different windows or glass doors have different *PF* values, they shall each be evaluated separately, or an area-weighted *PF* value shall be calculated and used for all windows and glass doors.

502.4 Air leakage control (Mandatory Requirements).

502.4.1 Window and door assemblies. The air leakage of window and sliding or swinging door assemblies that are part of the building envelope shall be determined in accordance with AAMA/WDMA/CSA 101/I.S.2/A440, or NFRC 400 by an accredited (by a body such as IAS or NFRC), independent laboratory, and *labeled* and certified by the manufacturer and shall not exceed the values in Section 402.4.4.

> **Exception:** Site-constructed windows and doors that are weatherstripped or sealed in accordance with Section 502.4.3.

502.4.2 Curtain wall, storefront glazing and commercial entrance doors. Curtain wall, *storefront* glazing and commercial-glazed swinging entrance doors and revolving doors shall be tested for air leakage at 1.57 pounds per square foot (psf) (75 Pa) in accordance with ASTM E 283. For curtain walls and *storefront* glazing, the maximum air leakage rate shall be 0.3 cubic foot per minute per square foot (cfm/ft²) (5.5 m³/h × m²) of fenestration area. For commercial glazed swinging entrance doors and revolving doors, the maximum air leakage rate shall be 1.00 cfm/ft² (18.3 m³/h × m²) of door area when tested in accordance with ASTM E 283.

TABLE 502.3
BUILDING ENVELOPE REQUIREMENTS: FENESTRATION

CLIMATE ZONE	3	4	5
Vertical fenestration (30% maximum of above-grade wall)			
U-factor			
Framing materials other than metal with or without metal reinforcement or cladding			
U-factor	0.32	0.32	0.30
Metal framing with or without thermal break			
Curtain wall/storefront *U*-factor	0.45	0.45	0.38
Entrance door *U*-factor	0.77	0.77	0.77
All other *U*-factor[a]	0.45	0.45	0.45
SHGC-all frame types			
SHGC: PF < 0.25	0.25	0.25	0.40
SHGC: 0.25 ≤ PF < 0.5	0.33	0.33	NR
SHGC: PF ≥ 0.5	0.40	0.40	NR
Skylights (3% maximum, 5% if using automatic daylighting controls)			
U-factor	0.60	0.60	0.60
SHGC	0.35	0.35	0.40

SHGC = Solar Heat Gain Coefficient.(approximately equal to 0.87 times the Shading Coefficient).

NR = No requirement.

PF = Projection factor (see Section 502.3.2).

a. All others includes operable windows, fixed windows and nonentrance doors.

502.4.3 Sealing of the building envelope. Openings and penetrations in the building envelope shall be sealed with caulking materials or closed with gasketing systems compatible with the construction materials and location. Joints and seams of the air barrier system shall be sealed in the same manner or taped or covered with a moisture vapor-permeable wrapping material. Sealing materials spanning joints between construction materials shall allow for expansion and contraction of the construction materials. See construction details in Appendix 2.1.

The following connections shall be air sealed:

1. Joints around fenestration and door frames

2. Junctions between walls and foundations, between walls at building corners, between walls and structural floors or roofs, and between walls and roof or wall panels

3. Openings at penetrations of utility services through roofs, walls, and floors including but not limited to electrical, plumbing, mechanical, security, and communications

4. Site-built fenestration and doors

5. Joints, seams, and penetrations of the air barrier system

6. Other openings in the building envelope

502.4.5 Stair and elevator vents. Shaft vents serving stairs and elevators integral to the building thermal envelope shall be equipped with not less than a Class I motorized, leakage-rated damper with a maximum leakage rate of 4 cfm per square foot (6.8 L/s · Cm^2) at 1.0 inch water gauge (w.g.) (125 Pa) when tested in accordance with AMCA 500D.

Exceptions:

1. Buildings without fire alarm systems.

2. Stairway vents open to the exterior.

502.4.6 Loading dock weatherseals. Cargo doors and loading dock doors shall be equipped with weatherseals to restrict infiltration when vehicles are parked in the doorway.

502.4.7 Vestibules. A door that separates *conditioned space* from the exterior shall be protected with an enclosed vestibule, with all doors opening into and out of the vestibule equipped with self-closing devices. Vestibules shall be designed so that in passing through the vestibule it is not necessary for the interior and exterior doors to open at the same time.

Exceptions:

1. Doors not intended to be used as a building *entrance door*, such as doors to mechanical or electrical equipment rooms.

2. Doors opening directly from a *sleeping unit* or dwelling unit.

3. Doors that open directly from a space less than 3,000 square feet (298 m²) in area.

4. Revolving doors.

5. Doors used primarily to facilitate vehicular movement or material handling and adjacent personnel doors.

6. Building entrances in buildings that are less than four stories above grade and less than 10,000 square feet in area.

502.4.8 Recessed lighting. Recessed luminaires installed in the *building thermal envelope* shall be IC-rated and labeled as meeting ASTM E 283 when tested at 1.57 psf (75 Pa) pressure differential with no more than 2.0 cfm (0.944 L/s) of air movement from the *conditioned space* to the ceiling cavity. Such recessed luminaires shall be sealed with a gasket or caulk between the housing and interior wall or ceiling covering.

SECTION 503
BUILDING MECHANICAL SYSTEMS

503.1 General. Mechanical systems and equipment serving the building heating, cooling or ventilating needs shall comply with Section 503.2 (referred to as the mandatory provisions) and either:

1. Section 503.3 (Simple systems), or

2. Section 503.4 (Complex systems).

503.2 Provisions applicable to all mechanical systems (Mandatory requirements).

503.2.1 Calculation of heating and cooling loads. Design loads shall be determined in accordance with the procedures described in the ASHRAE/ACCA Standard 183. Heating and cooling loads shall be adjusted to account for load reductions that are achieved when energy recovery systems are utilized in the HVAC system in accordance with the ASHRAE *HVAC Systems and Equipment Handbook*. Alternatively, design loads shall be determined by an *approved* equivalent computation procedure, using the design parameters specified in Chapter 3.

503.2.2 Equipment and system sizing. Heating and cooling equipment and systems capacity shall not exceed the loads calculated in accordance with Section 503.2.1, within available equipment options. A single piece of equipment providing both heating and cooling must satisfy this provision for one function with the capacity for the other function as small as possible.

Exceptions:

1. Required standby equipment and systems provided with controls and devices that allow such systems or equipment to operate automatically only when the primary equipment is not operating.

2. Multiple units of the same equipment type with combined capacities exceeding the design load and provided with controls that have the capability to sequence the operation of each unit based on load.

3. When the equipment selected is the smallest size needed to meet the load within available options of the desired equipment line.

503.2.3 HVAC equipment performance requirements. Equipment shall meet the minimum efficiency requirements of Tables 503.2.3(1), 503.2.3(2), 503.2.3(3), 503.2.3(4), 503.2.3(5), 503.2.3(6) and 503.2.3(7) when tested and rated in accordance with the applicable test procedure. The efficiency shall be verified through certification under an *approved* certification program or, if no certification program exists, the equipment efficiency ratings shall be supported by data furnished by the manufacturer. Where multiple rating conditions or performance requirements are provided, the equipment shall satisfy all stated requirements. Where components, such as indoor or outdoor coils, from different manufacturers are used, calculations and supporting data shall be furnished by the designer that demonstrates that the combined efficiency of the specified components meets the requirements herein.

Exception: Water-cooled centrifugal water-chilling packages listed in Table 503.2.3(7) not designed for operation at ARHI Standard 550/590 test conditions of 44°F (7°C) leaving chilled water temperature and 85°F (29°C) entering condenser water temperature with 3 gpm/ton (0.054 I/s.kW) condenser water flow shall have maximum full load and NPLV ratings adjusted using the following equations:

Adjusted maximum full load kW/ton rating =
[full load kW/ton from Table 503.2.3(7)]/K_{adj}

Adjusted maximum NPLV rating =
[IPLV from Table 503.2.3(7)]/K_{adj}

where:

K_{adj} = $6.174722 - 0.303668(X) + 0.00629466(X)^2 - 0.000045780(X)^3$

X = DT_{std} + LIFT

DT_{std} = {24 + [full load kW/ton from Table 503.2.3(7)] × 6.83}/Flow

Flow = Condenser water flow (GPM)/Cooling Full Load Capacity (tons)

LIFT = CEWT − CLWT (°F)

CEWT = Full Load Condenser Entering Water Temperature (°F)

CLWT = Full Load Leaving Chilled Water Temperature (°F)

The adjusted full load and NPLV values are only applicable over the following full-load design ranges:

Minimum Leaving Chilled Water Temperature: 38°F (3.3°C)

Maximum Condenser Entering Water Temperature: 102°F (38.9°C)

Condensing Water Flow: 1 to 6 gpm/ton (0.018 to 0.1076 1/s · kW) and X ≥ 39 and ≤ 60

Chillers designed to operate outside of these ranges or applications utilizing fluids or solutions with secondary coolants (e.g., glycol solutions or brines) with a freeze point of 27°F (-2.8°C) or lower for freeze protection are not covered by this code.

503.2.4 HVAC system controls. Each heating and cooling system shall be provided with thermostatic controls as required in Section 503.2.4.1, 503.2.4.2, 503.2.4.3, 503.2.4.4, 503.4.1, 503.4.2, 503.4.3 or 503.4.4.

503.2.4.1 Thermostatic controls. The supply of heating and cooling energy to each zone shall be controlled by individual thermostatic controls capable of responding to temperature within the zone. Where humidification or dehumidification or both is provided, at least one humidity control device shall be provided for each humidity control system.

Exception: Independent perimeter systems that are designed to offset only building envelope heat losses or gains or both serving one or more perimeter zones also served by an interior system provided:

1. The perimeter system includes at least one thermostatic control zone for each building exposure having exterior walls facing only one orientation (within +/- 45 degrees) (0.8 rad) for more than 50 contiguous feet (15.2 m); and

2. The perimeter system heating and cooling supply is controlled by a thermostat(s) located within the zone(s) served by the system.

TABLE 503.2.3(1)
UNITARY AIR CONDITIONERS AND CONDENSING UNITS, ELECTRICALLY OPERATED, MINIMUM EFFICIENCY REQUIREMENTS

EQUIPMENT TYPE	SIZE CATEGORY[e]	SUBCATEGORY OR RATING CONDITION	MINIMUM EFFICIENCY[b]	TEST PROCEDURE[a]
Air conditioners, Air cooled	< 65,000 Btu/h[d] (≤ 5 nominal tons)	Split system	13.0 SEER	AHRI 210/240
		Single package	13.0 SEER	
	≥ 65,000 Btu/h and < 135,000 Btu/h	Split system and single package	11.2 EER[c]	
	≥ 135,000 Btu/h and < 240,000 Btu/h	Split system and single package	11.0 EER[c]	AHRI 340/360
	≥ 240,000 Btu/h and < 760,000 Btu/h	Split system and single package	10.0 EER[c] 9.7 IPLV[g]	
	≥ 760,000 Btu/h	Split system and single package	9.7 EER[c] 9.4 IPLV[c]	
Through-the-wall, Air cooled	< 30,000 Btu/h[d]	Split system	12.0 SEER	AHRI 210/240
		Single package	12.0 SEER	
Air conditioners, Water and evaporatively cooled	< 65,000 Btu/h	Split system and single package	12.1 EER	AHRI 210/240
	≥ 65,000 Btu/h and < 135,000 Btu/h	Split system and single package	11.5 EER[c]	
	≥ 135,000 Btu/h and < 240,000 Btu/h	Split system and single package	11.0 EER[c]	AHRI 340/360
	≥ 240,000 Btu/h	Split system and single package	11.5 EER[c]	

For SI: 1 British thermal unit per hour = 0.2931 W.

a. Chapter 6 contains a complete specification of the referenced test procedure, including the referenced year version of the test procedure.

b. IPLVs are only applicable to equipment with capacity modulation.

c. Deduct 0.2 from the required EERs and IPLVs for units with a heating section other than electric resistance heat.

d. Single-phase air-cooled air conditioners < 65,000 Btu/h are regulated by the National Appliance Energy Conservation Act of 1987 (NAECA); SEER values are those set by NAECA.

e. Size category is based on nominal equipment sizes, e.g. < 65,000 Btu/h indicates ≤ 5 tons, < 135,000 Btu/h indicates ≤ 10 tons, < 240,000 Btu/h indicates < 20 tons, ≥ 760,000 Btu/h indicates > 60 tons.

TABLE 503.2.3(2)
UNITARY AIR CONDITIONERS AND CONDENSING UNITS, ELECTRICALLY OPERATED, MINIMUM EFFICIENCY REQUIREMENTS

EQUIPMENT TYPE	SIZE CATEGORY[e]	SUBCATEGORY OR RATING CONDITION	MINIMUM EFFICIENCY[b]	TEST PROCEDURE[a]
Air cooled, (Cooling mode)	< 65,000 Btu/h[d]	Split system	13.0 SEER	AHRI 210/240
		Single package	13.0 SEER	
	≥ 65,000 Btu/h and < 135,000 Btu/h	Split system and single package	11.0 EER[c]	
	≥ 135,000 Btu/h and < 240,000 Btu/h	Split system and single package	10.6 EER[c]	AHRI 340/360
	≥ 240,000 Btu/h	Split system and single package	9.5 EER[c] 9.2 IPLV[c]	
Through-the-Wall (Air cooled, cooling mode)	< 30,000 Btu/h[d]	Split system	12.0 SEER	AHRI 210/240
		Single package	12.0 SEER	
Water Source (Cooling mode)	< 17,000 Btu/h	86°F entering water	11.2 EER	AHRI/ASHRAE 13256-1
	≥ 17,000 Btu/h and < 135,000 Btu/h	86°F entering water	12.0 EER	AHRIASHRAE 13256-1
Groundwater Source (Cooling mode)	< 135,000 Btu/h	59°F entering water	16.2 EER	AHRI/ASHRAE 13256-1
Ground source (Cooling mode)	< 135,000 Btu/h	77°F entering water	13.4 EER	AHRI/ASHRAE 13256-1
Air cooled (Heating mode)	< 65,000 Btu/h[d] (Cooling capacity)	Split system	7.7 HSPF	AHRI 210/240
		Single package	7.7 HSPF	
	≥ 65,000 Btu/h and < 135,000 Btu/h (Cooling capacity)	47°F db/43°F wb outdoor air	3.3 COP	
	≥ 135,000 Btu/h (Cooling capacity)	47°F db/43°F wb outdoor air	3.2 COP	AHRI 340/360
Through-the-wall (Air cooled, heating mode)	< 30,000 Btu/h	Split system	7.4 HSPF	AHRI 210/240
		Single package	7.4 HSPF	
Water source (Heating mode)	< 135,000 Btu/h (Cooling capacity)	68°F entering water	4.2 COP	AHRI/ASHRAE 13256-1
Groundwater source (Heating mode)	< 135,000 Btu/h (Cooling capacity)	50°F entering water	3.6 COP	AHRI/ASHRAE 13256-1
Ground source (Heating mode)	< 135,000 Btu/h (Cooling capacity)	32°F entering water	3.1 COP	AHRI/ASHRAE 13256-1

For SI: °C = [(°F) - 32]/1.8, 1 British thermal unit per hour = 0.2931 W.

db = dry-bulb temperature, °F; wb = wet-bulb temperature, °F.

a. Chapter 6 contains a complete specification of the referenced test procedure, including the referenced year version of the test procedure.

b. IPLVs and Part load rating conditions are only applicable to equipment with capacity modulation.

c. Deduct 0.2 from the required EERs and IPLVs for units with a heating section other than electric resistance heat.

d. Single-phase air-cooled heat pumps < 65,000 Btu/h are regulated by the National Appliance Energy Conservation Act of 1987 (NAECA), SEER and HSPF values are those set by NAECA.

e. Size category is based on nominal equipment sizes, e.g. < 65,000 Btu/h indicates ≤ 5 tons, < 135,000 Btu/h indicates ≤ 10 tons, < 240,000 Btu/h indicates < 20 tons, ≥ 760,000 Btu/h indicates > 60 tons.

TABLE 503.2.3(3)
PACKAGED TERMINAL AIR CONDITIONERS AND PACKAGED TERMINAL HEAT PUMPS

EQUIPMENT TYPE	SIZE CATEGORY (INPUT)	SUBCATEGORY OR RATING CONDITION	MINIMUM EFFICIENCY[b]	TEST PROCEDURE[a]
PTAC (Cooling mode) New construction	All capacities	95°F db outdoor air	12.5 - (0.213 · Cap/1000) EER	AHRI 310/380
PTAC (Cooling mode) Replacements[c]	All capacities	95°F db outdoor air	10.9 - (0.213 · Cap/1000) EER	
PTHP (Cooling mode) New construction	All capacities	95°F db outdoor air	12.3 - (0.213 · Cap/1000) EER	
PTHP (Cooling mode) Replacements[c]	All capacities	95°F db outdoor air	10.8 - (0.213 · Cap/1000) EER	
PTHP (Heating mode) New construction	All capacities	—	3.2 - (0.026 · Cap/1000) COP	
PTHP (Heating mode) Replacements[c]	All capacities	—	2.9 - (0.026 · Cap/1000) COP	

For SI: °C - [(°F) - 32]/1.8, 1 British thermal unit per hour - 0.2931 W.

db = dry-bulb temperature, °F.

wb = wet-bulb temperature, °F.

a. Chapter 6 contains a complete specification of the referenced test procedure, including the referenced year version of the test procedure.

b. Cap means the rated cooling capacity of the product in Btu/h. If the unit's capacity is less than 7,000 Btu/h, use 7,000 Btu/h in the calculation. If the unit's capacity is greater than 15,000 Btu/h, use 15,000 Btu/h in the calculation.

c. Replacement units must be factory labeled as follows: "MANUFACTURED FOR REPLACEMENT APPLICATIONS ONLY: NOT TO BE INSTALLED IN NEW CONSTRUCTION PROJECTS." Replacement efficiencies apply only to units with existing sleeves less than 16 inches (406 mm) high and less than 42 inches (1067 mm) wide.

TABLE 503.2.3(4)
WARM AIR FURNACES AND COMBINATION WARM AIR FURNACES/AIR-CONDITIONING UNITS,
WARM AIR DUCT FURNACES AND UNIT HEATERS, MINIMUM EFFICIENCY REQUIREMENTS

EQUIPMENT TYPE	SIZE CATEGORY (INPUT)[h]	SUBCATEGORY OR RATING CONDITION	MINIMUM EFFICIENCY[d, e]	TEST PROCEDURE[a]
Warm air furnaces, gas fired	< 225,000 Btu/h	—	78% AFUE or 80% E_t[c]	DOE 10 CFR Part 430 or ANSI Z21.47
	≥ 225,000 Btu/h	Maximum capacity[c]	80% E_t[f]	ANSI Z21.47
Warm air furnaces, oil fired	< 225,000 Btu/h	—	78% AFUE or 80% E_t[c]	DOE 10 CFR Part 430 or UL 727
	≥ 225,000 Btu/h	Maximum capacity[b]	81% E_t[g]	UL 727
Warm air duct furnaces, gas fired	All capacities	Maximum capacity[b]	80% E_c	ANSI Z83.8
Warm air unit heaters, gas fired	All capacities	Maximum capacity[b]	80% E_c	ANSI Z83.8
Warm air unit heaters, oil fired	All capacities	Maximum capacity[b]	80% E_c	UL 731

For SI: 1 British thermal unit per hour = 0.2931 W.

a. Chapter 6 contains a complete specification of the referenced test procedure, including the referenced year version of the test procedure.

b. Minimum and maximum ratings as provided for and allowed by the unit's controls.

c. Combination units not covered by the National Appliance Energy Conservation Act of 1987 (NAECA) (3-phase power or cooling capacity greater than or equal to 65,000 Btu/h [19 kW]) shall comply with either rating.

d. E_t = Thermal efficiency. See test procedure for detailed discussion.

e. E_c = Combustion efficiency (100% less flue losses). See test procedure for detailed discussion.

f. E_c = Combustion efficiency. Units must also include an IID, have jackets not exceeding 0.75 percent of the input rating, and have either power venting or a flue damper. A vent damper is an acceptable alternative to a flue damper for those furnaces where combustion air is drawn from the conditioned space.

g. E_t = Thermal efficiency. Units must also include an IID, have jacket losses not exceeding 0.75 percent of the input rating, and have either power venting or a flue damper. A vent damper is an acceptable alternative to a flue damper for those furnaces where combustion air is drawn from the conditioned space.

h. Size category is based on nominal equipment sizes, e.g. < 65,000 Btu/h indicates ≤ 5 tons, < 135,000 Btu/h indicates ≤ 10 tons, < 240,000 Btu/h indicates < 20 tons, ≥ 760,000 Btu/h indicates > 60 tons.

TABLE 503.2.3(5)
BOILERS, GAS- AND OIL-FIRED, MINIMUM EFFICIENCY REQUIREMENTS

EQUIPMENT TYPE[f]	SIZE CATEGORY (input)[g]	SUBCATEGORY OR RATING CONDITION	MINIMUM EFFICIENCY[c, d, e]	TEST PROCEDURE[a]
Boilers, Gas-fired	< 300,000 Btu/h	Hot water	80% AFUE	DOE 10 CFR Part 430
		Steam	75% AFUE	
	≥ 300,000 Btu/h and ≤ 2,500,000 Btu/h	Minimum capacity[b]	75% E_t and 80% E_c (See Note c, d)	DOE 10 CFR Part 431
	> 2,500,000 Btu/h[f]	Hot water	80% E_c (See Note c, d)	
		Steam	80% E_c (See Note c, d)	
Boilers, Oil-fired	< 300,000 Btu/h	—	80% AFUE	DOE 10 CFR Part 430
	≥ 300,000 Btu/h and ≤ 2,500,000 Btu/h	Minimum capacity[b]	78% E_t and 83% E_c (See Note c, d)	DOE 10 CFR Part 431
	> 2,500,000 Btu/h[f]	Hot water	83% E_c (See Note c, d)	
		Steam	83% E_c (See Note c, d)	
Boilers, Oil-fired (Residual)	≥ 300,000 Btu/h and ≤ 2,500,000 Btu/h	Minimum capacity[b]	78% E_t and 83% E_c (See Note c, d)	DOE 10 CFR Part 431
	> 2,500,000 Btu/h[f]	Hot water	83% E_c (See Note c, d)	
		Steam	83% E_c (See Note c, d)	

For SI: 1 British thermal unit per hour = 0.2931 W.

a. Chapter 6 contains a complete specification of the referenced test procedure, including the referenced year version of the test procedure.

b. Minimum ratings as provided for and allowed by the unit's controls.

c. E_c = Combustion efficiency (100 percent less flue losses). See reference document for detailed information.

d. E_t = Thermal efficiency. See reference document for detailed information.

e. Alternative test procedures used at the manufacturer's option are ASME PTC-4.1 for units greater than 5,000,000 Btu/h input, or ANSI Z21.13 for units greater than or equal to 300,000 Btu/h and less than or equal to 2,500,000 Btu/h input.

f. These requirements apply to boilers with rated input of 8,000,000 Btu/h or less that are not packaged boilers, and to all packaged boilers. Minimum efficiency requirements for boilers cover all capacities of packaged boilers.

g. Size category is based on nominal equipment sizes, e.g. < 65,000 Btu/h indicates ≤ 5 tons, < 135,000 Btu/h indicates ≤ 10 tons, < 240,000 Btu/h indicates < 20 tons, ≥ 760,000 Btu/h indicates > 60 tons.

TABLE 503.2.3(6)
CONDENSING UNITS, ELECTRICALLY OPERATED, MINIMUM EFFICIENCY REQUIREMENTS

EQUIPMENT TYPE	SIZE CATEGORY	MINIMUM EFFICIENCY[b]	TEST PROCEDURE[a]
Condensing units, air cooled	≥ 135,000 Btu/h	10.1 EER 11.2 IPLV	AHRI 365
Condensing units, water or evaporatively cooled	≥ 135,000 Btu/h	13.1 EER 13.1 IPLV	

For SI: 1 British thermal unit per hour = 0.2931 W.

a. Chapter 6 contains a complete specification of the referenced test procedure, including the referenced year version of the test procedure.

b. IPLVs are only applicable to equipment with capacity modulation.

TABLE 503.2.3(7)
WATER CHILLING PACKAGES, EFFICIENCY REQUIREMENTS[a]

| EQUIPMENT TYPE | SIZE CATEGORY | UNITS | BEFORE 1/1/2010 | | AS OF 1/1/2010[c] | | | | TEST PROCEDURE[b] |
| | | | FULL LOAD | IPLV | PATH A | | PATH B | | |
					FULL LOAD	IPLV	FULL LOAD	IPLV	
Air-cooled chillers	< 150 tons	EER	≥ 9.562	≥ 10.416	≥ 9.562	≥ 12.500	NA[d]	NA[d]	AHRI 550/590
	≥ 150 tons	EER			≥ 9.562	≥ 12.750	NA[d]	NA[d]	
Air cooled without condenser, electrical operated	All capacities	EER	≥ 10.586	≥ 11.782	Air-cooled chillers without condensers must be rated with matching condensers and comply with the air-cooled chiller efficiency requirements				
Water cooled, electrically operated, reciprocating	All capacities	kW/ton	≤ 0.837	≤ 0.696	Reciprocating units must comply with water cooled positive displacement efficiency requirements				
Water cooled, electrically operated, positive displacement	< 75 tons	kW/ton	≤ 0.790	≤ 0.676	≤ 0.780	≤ 0.630	≤ 0.800	≤ 0.600	
	≥ 75 tons and < 150 tons	kW/ton			≤ 0.775	≤ 0.615	≤ 0.790	≤ 0.586	
	≥ 150 tons and < 300 tons	kW/ton	≤ 0.717	≤ 0.627	≤ 0.680	≤ 0.580	≤ 0.718	≤ 0.540	
	≥ 300 tons	kW/ton	≤ 0.639	≤ 0.571	≤ 0.620	≤ 0.540	≤ 0.639	≤ 0.490	
Water cooled, electrically operated, centrifugal	< 150 tons	kW/ton	≤ 0.703	≤ 0.669	≤ 0.634	≤ 0.596	≤ 0.639	≤ 0.450	
	≥ 150 tons and < 300 tons	kW/ton	≤ 0.634	≤ 0.596					
	≥ 300 tons and < 600 tons	kW/ton	≤ 0.576	≤ 0.549	≤ 0.576	≤ 0.549	≤ 0.600	≤ 0.400	
	≥ 600 tons	kW/ton	≤ 0.576	≤ 0.549	≤ 0.570	≤ 0.539	≤ 0.590	≤ 0.400	
Air cooled, absorption single effect	All capacities	COP	≥ 0.600	NR[e]	≥ 0.600	NR[e]	NA[d]	NA[d]	AHRI 560
Water-cooled, absorption single effect	All capacities	COP	≥ 0.700	NR[e]	≥ 0.700	NR[e]	NA[d]	NA[d]	
Absorption double effect, indirect-fired	All capacities	COP	≥ 1.000	≥ 1.050	≥ 1.000	≥ 1.050	NA[d]	NA[d]	
Absorption double effect, direct fired	All capacities	COP	≥ 1.000	≥ 1.000	≥ 1.000	≥ 1.000	NA[d]	NA[d]	

For SI: 1 ton = 3517 W, 1 British thermal unit per hour = 0.2931 W.

a. The chiller equipment requirements do not apply for chillers used in low-temperature applications where the design leaving fluid temperature is < 40°F.

b. Section 12 contains a complete specification of the referenced test procedure, including the referenced year version of the test procedure.

c. Compliance with this standard can be obtained by meeting the minimum requirements of Path A or B. However, both the full load and IPLV must be met to fulfill the requirements of Path A or B.

d. NA means that this requirement is not applicable and cannot be used for compliance.

e. NR means that there are no minimum requirements for this category.

503.2.4.1.1 Heat pump supplementary heat. Heat pumps having supplementary electric resistance heat shall have controls that, except during defrost, prevent supplementary heat operation when the heat pump can meet the heating load.

In systems with a cooling capacity of less than 65,000 Btuh, a heat strip outdoor temperature lockout shall be provided to prevent supplemental heat operation in response to the thermostat being changed to a warmer setting. The lockout shall be set no lower than 35°F and no higher than 40°F.

503.2.4.2 Set point overlap restriction. Where used to control both heating and cooling, zone thermostatic controls shall provide a temperature range or deadband of at least 5°F (2.8°C) within which the supply of heating and cooling energy to the zone is capable of being shut off or reduced to a minimum.

Exception: Thermostats requiring manual change-over between heating and cooling modes.

503.2.4.3 Off-hour controls. Each zone shall be provided with thermostatic setback controls that are controlled by either an automatic time clock or programmable control system.

Exceptions:

1. Zones that will be operated continuously.

2. Zones with a full HVAC load demand not exceeding 6,800 Btu/h (2 kW) and having a readily accessible manual shutoff switch.

3. HVAC systems serving hotel/motel guestrooms or other residential units complying with Section 503.2.2 requirements.

4. Packaged terminal air conditioners, packaged terminal heat pumps, and room air conditioner systems.

503.2.4.3.1 Thermostatic setback capabilities. Thermostatic setback controls shall have the capability to set back or temporarily operate the system to maintain zone temperatures down to 55°F (13°C) or up to 85°F (29°C).

503.2.4.3.2 Automatic setback and shutdown capabilities. Automatic time clock or programmable controls shall be capable of starting and stopping the system for seven different daily schedules per week and retaining their programming and time setting during a loss of power for at least 10 hours. Additionally, the controls shall have a manual override that allows temporary operation of the system for up to 2 hours; a manually operated timer capable of being adjusted to operate the system for up to 2 hours; or an occupancy sensor.

503.2.4.4 Shutoff damper controls. Outdoor air supply and exhaust ducts, fans or openings in the building thermal envelope shall be equipped with motorized dampers that will automatically shut when the systems or spaces served are not in use.

Exceptions:

1. Gravity dampers shall be permitted in buildings less than three stories in height.

2. Gravity dampers shall be permitted for outside air intake or exhaust airflows of 300 cfm (0.14m³/s) or less.

503.2.4.5 Snow melt system controls. Snow- and ice-melting systems, supplied through energy service to the building, shall include automatic controls capable of shutting off the system when the pavement temperature is above 50°F (10°C) and no precipitation is falling and an automatic or manual control that will allow shutoff when the outdoor temperature is above 40°F (4°C) so that the potential for snow or ice accumulation is negligible.

503.2.5 Ventilation. Ventilation, either natural or mechanical, shall be provided in accordance with Chapter 4 of the *International Mechanical Code*. Where mechanical ventilation is provided, the system shall provide the capability to reduce the outdoor air supply to the minimum required by Chapter 4 of the *International Mechanical Code*.

503.2.5.1 Demand controlled ventilation. Ventilation systems in buildings over 10,000 square feet of conditioned area shall have demand controls. In all buildings, spaces larger than 500 square feet (50 m²) with a maximum occupant load of 40 or more people per 1,000 square feet (93 m²) of floor area (as established in Table 403.3 of the *North Carolina Mechanical Code*), ventilation supply air flow shall be controlled by monitoring indoor air quality conditions, such as with CO^2 sensors or thermostats. Demand controlled ventilation systems shall be capable of reducing outside supply air to at least 50 percent below design ventilation rates.

Exceptions:

1. Systems with energy recovery that provide a change in the enthalpy of the outdoor air supply of 50 percent or more of the difference between the outdoor air and return air at design conditions.

2. Building spaces where the primary ventilation needs are for process loads, including laboratories and hospital.

3. Individual units with less than 65,000 Btu/h of cooling capacity.

503.2.6 Energy recovery ventilation systems. Individual fan systems that have both a design supply air capacity of 5,000 cfm (2.36 m³/s) or greater and a minimum outside air supply of 70 percent or greater of the design supply air quantity shall have an energy recovery system that provides a change in the enthalpy of the outdoor air supply of 50 percent or more of the difference between the outdoor air and return air at design conditions. Provision shall be made to bypass or control the energy recovery system to permit cooling with outdoor air where cooling with outdoor air is required.

Exception: An energy recovery ventilation system shall not be required in any of the following conditions:

1. Where energy recovery systems are prohibited by the *International Mechanical Code*.

2. Laboratory fume hood systems that include at least one of the following features:

    2.1. Variable-air-volume hood exhaust and room supply systems capable of reducing exhaust and makeup air volume to 50 percent or less of design values.

    2.2. Direct makeup (auxiliary) air supply equal to at least 75 percent of the exhaust rate, heated no warmer than 2°F (1.1°C) below room setpoint, cooled to no cooler than 3°F (1.7°C) above room setpoint, no humidification added, and no simultaneous heating and cooling used for dehumidification control.

3. Systems serving spaces that are not cooled and are heated to less than 60°F (15.5°C).

4. Where more than 60 percent of the outdoor heating energy is provided from site-recovered or site solar energy.

5. Heating systems in climates with less than 3,600 HDD.

6. Cooling systems in climates with a 1-percent cooling design wet-bulb temperature less than 64°F (18°C).

7. Systems requiring dehumidification that employ series-style energy recovery coils wrapped around the cooling coil.

503.2.7 Duct and plenum insulation and sealing. All supply and return air ducts and plenums shall be insulated with a minimum of R-5 insulation when located in unconditioned spaces and a minimum of R-8 insulation when located outside the building. When located within a building envelope assembly, the duct or plenum shall be separated from the building exterior or unconditioned or exempt spaces by a minimum of R-8 insulation.

Exceptions:

1. When located within equipment.

2. When the design temperature difference between the interior and exterior of the duct or plenum does not exceed 15°F (8°C).

All ducts, air handlers and filter boxes shall be sealed. Joints and seams shall comply with Section 603.9 of the *International Mechanical Code*.

503.2.7.1 Duct construction. Ductwork shall be constructed and erected in accordance with the *International Mechanical Code*.

503.2.7.1.1 Low-pressure duct systems. All longitudinal and transverse joints, seams and connections of supply and return ducts operating at a static pressure less than or equal to 2 inches w.g. (500 Pa) shall be securely fastened and sealed with welds, gaskets, mastics (adhesives), mastic-plus-embedded-fabric systems or tapes installed in accordance with the manufacturer's installation instructions. Pressure classifications specific to the duct system shall be indicated on the construction documents in accordance with the *International Mechanical Code*.

Exceptions:

1. Continuously welded and locking-type longitudinal joints and seams on ducts operating at static pressures less than 2 inches w.g. (500 Pa) pressure classification.

2. Ducts exposed within the conditioned space they serve shall not be required to be sealed.

503.2.7.1.2 Medium-pressure duct systems. All ducts and plenums designed to operate at a static pressure greater than 2 inches w.g. (500 Pa) but less than 3 inches w.g. (750 Pa) shall be insulated and sealed in accordance with Section 503.2.7. Pressure classifications specific to the duct system shall be indicated on the construction documents in accordance with the *International Mechanical Code*.

503.2.7.1.3 High-pressure duct systems. Ducts designed to operate at static pressures in excess of 3 inches w.g. (746 Pa) shall be insulated and sealed in accordance with Section 503.2.7. In addition, ducts and plenums shall be leak-tested in accordance with the SMACNA *HVAC Air Duct Leakage Test Manual* with the rate of air leakage (CL) less than or equal to 6.0 as determined in accordance with Equation 5-2.

$$CL = F \times P^{0.65} \qquad \text{(Equation 5-2)}$$

where:

F = The measured leakage rate in cfm per 100 square feet of duct surface.

P = The static pressure of the test.

Documentation shall be furnished by the designer demonstrating that representative sections totaling at least 25 percent of the duct area have been tested and that all tested sections meet the requirements of this section.

503.2.8 Piping insulation. All piping serving as part of a heating or cooling system shall be thermally insulated in accordance with Table 503.2.8.

Exceptions:

1. Factory-installed piping within HVAC equipment tested and rated in accordance with a test procedure referenced by this code.

2. Factory-installed piping within room fan-coils and unit ventilators tested and rated according to AHRI 440 (except that the sampling and variation provisions of Section 6.5 shall not apply) and 840, respectively.

3. Piping that conveys fluids that have a design operating temperature range between 55°F (13°C) and 105°F (41°C).

4. Piping that conveys fluids that have not been heated or cooled through the use of fossil fuels or electric power.

5. Runout piping not exceeding 4 feet (1219 mm) in length and 1 inch (25 mm) in diameter between the control valve and HVAC coil.

6. Refrigerant suction piping located in conditioned space is not required to be insulated other than as may be necessary for preventing the formation of condensation.

TABLE 503.2.8
MINIMUM PIPE INSULATION
(thickness in inches)

FLUID	NOMINAL PIPE DIAMETER	
	≤ 1.5"	> 1.5"
Steam	$1^1/_2$	3
Hot water	$1^1/_2$	2
Chilled water, brine or refrigerant	$1^1/_2$	$1^1/_2$

For SI: 1 inch = 25.4 mm.

a. Based on insulation having a conductivity (k) not exceeding 0.27 Btu per inch/h · ft² · °F.

b. For insulation with a thermal conductivity not equal to 0.27 Btu · inch/h · ft² · °F at a mean temperature of 75°F, the minimum required pipe thickness is adjusted using the following equation;

$$T = r[(1+t/r)^{K/k}-1]$$

where:

T = Adjusted insulation thickness (in).

r = Actual pipe radius (in).

t = Insulation thickness from applicable cell in table (in).

K = New thermal conductivity at 75°F (Btu · in/hr · ft² · °F).

k = 0.27 Btu · in/hr · ft² · °F.

503.2.9 HVAC System Completion. Prior to the issuance of a certificate of occupancy, the following shall be completed.

Exception: A temporary certificate of occupancy shall be allowed to be issued when requested prior to the completion of this section.

503.2.9.1 System balancing. All HVAC systems shall be balanced by contractor. Test and balance activities shall include the following items:

503.2.9.1.1 Air systems balancing. Each supply air outlet and zone terminal device shall be equipped with means for air balancing in accordance with the requirements of Chapter 6 of the *North Carolina Mechanical Code*. Discharge dampers are prohibited on constant volume fans and variable volume fans with motors 10 hp (7.5 kW) and larger.

Exception: Fan with fan motors of 1 hp or less.

503.2.9.1.2 Hydronic systems balancing. Individual hydronic heating and cooling coils shall be equipped with means for balancing and pressure test connections. Hydronic systems shall be balanced in a manner to first minimize throttling losses, then the pump impeller shall be trimmed or pump speed shall be adjusted to meet design flow conditions. Each hydronic system shall have either the ability to measure pressure across the pump, or test ports at each side of each pump.

Exceptions:

1. Pumps with pump motors of 5 hp or less.

2. When throttling of an individual pump results in no greater than 5% of the nameplate horsepower draw above that required if the impeller were trimmed.

503.2.9.2 Manuals. An operating and maintenance manual shall be provided to the building owner by the contractor. The manual shall include the following:

1. Submittal data stating equipment model number and capacity (input and output) and selected options for each piece of equipment.

2. Manufacturer's operation manuals and maintenance manuals for each piece of equipment requiring maintenance, except equipment not furnished as part of the project. Required routine maintenance actions shall be identified.

3. Name and address of at least one service agency.

4. HVAC controls system maintenance and calibration information, including wiring diagrams, schematics, and control sequence descriptions. Desired or field-determined set points shall be permanently recorded on control drawings at control devices or, for digital control systems, in programming comments.

5. A complete narrative of how each system is intended to operate.

6. Names and addresses of designers of record, contractors, subcontractors and equipment suppliers.

503.2.9.3 System installation statement. A North Carolina licensed design professional shall prepare and sign the *Statement of Compliance – HVAC System Installation (Appendix 5)*. This statement shall be submitted to the code official and the facility owner.

Exception: The HVAC contractor will be allowed to prepare the *Statement of Compliance* when a building permit is issued for a project without the seal of a licensed design professional as allowed by an exception under NC State Building Administrative Code and Policies: Section 204.3.5.

503.2.9.3.1 Equipment. Equipment installation and operation shall be verified, to the extent feasible, to be in accordance with approved plans and specifications. Verification shall include demonstration of operation of components, systems and system-to-system interfacing relationships.

503.2.9.3.2 Controls. Controls installation and operation shall be verified, to the extent feasible, to be in accordance with approved plans and specifications. Verification shall include demonstration of operation of control devices, systems and system-to-system interfacing relationships. Control sequences shall be functionally verified, to the extent feasible, to demonstrate operation in accordance with the intent of the approval plans and specifications.

503.2.10 Air system design and control. Each HVAC system having a total fan system motor nameplate horsepower (hp) exceeding 5 horsepower (hp) shall meet the provisions of Sections 503.2.10.1 through 503.2.10.2.

503.2.10.1 Allowable fan horsepower. Each HVAC system at fan system design conditions shall not exceed the allowable fan system motor nameplate hp (Option 1) or fan system bhp (Option 2) as shown in Table 503.2.10.1(1). This includes supply fans, return/relief fans, and fan-powered terminal units associated with systems providing heating or cooling capability.

Exceptions:

1. Hospital and laboratory systems that utilize flow control devices on exhaust or return to maintain space pressure relationships necessary for occupant health and safety or environmental control shall be permitted to use variable volume fan power limitation.

2. Individual exhaust fans with motor nameplate horsepower of 1 hp or less.

3. Fans exhausting air from fume hoods. (Note: If this exception is taken, no related exhaust side credits shall be taken from Table 503.2.10.1(2) and the Fume Exhaust Exception Deduction must be taken from Table 503.2.10.1(2).

503.2.10.2 Motor nameplate horsepower. For each fan, the selected fan motor shall be no larger than the first available motor size greater than the brake horsepower (bhp). The fan brake horsepower (bhp) shall be indicated on the design documents to allow for compliance verification by the *code official.*

Exceptions:

1. For fans less than 6 bhp, where the first available motor larger than the brake horsepower has a nameplate rating within 50 percent of the bhp, selection of the next larger nameplate motor size is allowed.

2. For fans 6 bhp and larger, where the first available motor larger than the bhp has a nameplate rating within 30 percent of the bhp, selection of the next larger nameplate motor size is allowed.

503.2.11 Heating outside a building. Systems installed to provide heat outside a building shall be radiant systems. Such heating systems shall be controlled by an occupancy sensing device or a timer switch, so that the system is automatically deenergized when no occupants are present.

503.3 Simple HVAC systems and equipment (Prescriptive). This section applies to buildings served by unitary or packaged HVAC equipment listed in Tables 503.2.3(1) through 503.2.3(5), each serving one zone and controlled by a single thermostat in the zone served. It also applies to two-pipe heating systems serving one or more zones, where no cooling system is installed.

This section does not apply to fan systems serving multiple zones, nonunitary or nonpackaged HVAC equipment and systems or hydronic or steam heating and hydronic cooling equipment and distribution systems that provide cooling or cooling and heating which are covered by Section 503.4.

TABLE 503.2.10.1(1)
FAN POWER LIMITATION

	LIMIT	CONSTANT VOLUME	VARIABLE VOLUME
Option 1: Fan system motor nameplate hp	Allowable nameplate motor hp	$hp \leq CFM_S \times 0.0011$	$hp \leq CFM_S \times 0.0015$
Option 2: Fan system bhp	Allowable fan system bhp	$bhp \leq CFM_S \times 0.00094 + A$	$bhp \leq CFM_S \times 0.0013 + A$

where:

CFM_S = The maximum design supply airflow rate to conditioned spaces served by the system in cubic feet per minute.

hp = The maximum combined motor nameplate horsepower.

Bhp = The maximum combined fan brake horsepower.

A = Sum of $[PD \times CFM_D / 4131]$.

where:

PD = Each applicable pressure drop adjustment from Table 503.2.10.1(2) in. w.c.

CFM_D = The design airflow through each applicable device from Table 503.2.10.1(2) in cubic feet per minute.

TABLE 503.2.10.1(2)
FAN POWER LIMITATION PRESSURE DROP ADJUSTMENT

DEVICE	ADJUSTMENT
Credits	
Fully ducted return and/or exhaust air systems with capacity ≤ 20,000 cfm	0.5 in w.c.
Fully ducted return and/or exhaust air systems with capacity > 20,000 cfm	2.0 in w.c.
Return and/or exhaust airflow control devices	0.5 in w.c
Exhaust filters, scrubbers or other exhaust treatment	The pressure drop of device calculated at fan system design condition.
Particulate filtration credit: MERV 9 thru 12	0.5 in w.c.
Particulate filtration credit: MERV 13 thru 15	0.9 in w.c.
Particulate filtration credit: MERV 16 and greater and electronically enhanced filters	Pressure drop calculated at 2x clean filter pressure drop at fan system design condition.
Carbon and other gas-phase air cleaners	Clean filter pressure drop at fan system design condition.
Heat recovery device	Pressure drop of device at fan system design condition.
Evaporative humidifier/cooler in series with another cooling coil	Pressure drop of device at fan system design conditions.
Sound attenuation section	0.15 in w.c.
Deductions	
Fume hood exhaust exception (required if Section 503.2.10.1.1, Exception 3, is taken)	-1.0 in w.c.

503.3.1 Economizers. Supply air economizers shall be provided on each cooling system as shown in Table 503.3.1(1).

Economizers shall be capable of providing 100-percent outdoor air, even if additional mechanical cooling is required to meet the cooling load of the building. Systems shall provide a means to relieve excess outdoor air during economizer operation to prevent overpressurizing the building. The relief air outlet shall be located to avoid recirculation into the building. Where a single room or space is supplied by multiple air systems, the aggregate capacity of those systems shall be used in applying this requirement.

Exceptions:

1. Where the cooling equipment is covered by the minimum efficiency requirements of Table 503.2.3(1) or 503.2.3(2) and meets or exceeds the minimum cooling efficiency requirement (EER) by the percentages shown in Table 503.3.1(2).

2. Systems with air or evaporatively cooled condensors and which serve spaces with open case refrigeration or that require filtration equipment in order to meet the minimum ventilation requirements of Chapter 4 of the *International Mechanical Code.*

503.3.2 Hydronic system controls. Hydronic systems of at least 300,000 Btu/h (87,930 W) design output capacity supplying heated and chilled water to comfort conditioning systems shall include controls that meet the requirements of Section 503.4.3.

TABLE 503.3.1(1)
ECONOMIZER REQUIREMENTS

CLIMATE ZONES	ECONOMIZER REQUIREMENT
1A, 1B, 2A, 7, 8	No requirement
2B, 3A, 3B, 3C, 4A, 4B, 4C, 5A, 5B, 5C, 6A, 6B	Economizers on all cooling systems ≥ 65,000Btu/h[a]

For SI: 1 British thermal unit per hour = 0.293 W.

a. The total capacity of all systems without economizers shall not exceed 480,000 Btu/h per building, or 20 percent of its air economizer capacity, whichever is greater.

TABLE 503.3.1(2)
EQUIPMENT EFFICIENCY PERFORMANCE
EXCEPTION FOR ECONOMIZERS

CLIMATE ZONES	COOLING EQUIPMENT PERFORMANCE IMPROVEMENT (EER OR IPLV)
3B	15% Efficiency Improvement
4B	20% Efficiency Improvement

503.4 Complex HVAC systems and equipment. (Prescriptive). This section applies to buildings served by HVAC equipment and systems not covered in Section 503.3.

503.4.1 Economizers. Supply air economizers shall be provided on each cooling system according to Table 503.3.1(1). Economizers shall be capable of operating at 100 percent outside air, even if additional mechanical cooling is required to meet the cooling load of the building.

Exceptions:

1. Systems utilizing water economizers that are capable of cooling supply air by direct or indirect evaporation or both and providing 100 percent of the

expected system cooling load at outside air temperatures of 50°F (10°C) dry bulb/45°F (7°C) wet bulb and below.

2. Where the cooling equipment is covered by the minimum efficiency requirements of Table 503.2.3(1), 503.2.3(2), or 503.2.3(6) and meets or exceeds the minimum EER by the percentages shown in Table 503.3.1(2).

3. Where the cooling equipment is covered by the minimum efficiency requirements of Table 503.2.3(7) and meets or exceeds the minimum integrated part load value (IPLV) by the percentages shown in Table 503.3.1(2).

503.4.2 Variable air volume (VAV) fan control. Individual VAV fans with motors of 10 horsepower (7.5 kW) or greater shall be:

1. Driven by a mechanical or electrical variable speed drive; or

2. The fan motor shall have controls or devices that will result in fan motor demand of no more than 30 percent of their design wattage at 50 percent of design airflow when static pressure set point equals one-third of the total design static pressure, based on manufacturer's certified fan data.

For systems with direct digital control of individual *zone* boxes reporting to the central control panel, the static pressure set point shall be reset based on the *zone* requiring the most pressure, i.e., the set point is reset lower until one *zone* damper is nearly wide open.

503.4.3 Hydronic systems controls. The heating of fluids that have been previously mechanically cooled and the cooling of fluids that have been previously mechanically heated shall be limited in accordance with Sections 503.4.3.1 through 503.4.3.3. Hydronic heating systems comprised of multiple-packaged boilers and designed to deliver conditioned water or steam into a common distribution system shall include automatic controls capable of sequencing operation of the boilers. Hydronic heating systems comprised of a single boiler and greater than 500,000 Btu/h input design capacity shall include either a multistaged or modulating burner.

503.4.3.1 Three-pipe system. Hydronic systems that use a common return system for both hot water and chilled water are prohibited.

503.4.3.2 Two-pipe changeover system. Systems that use a common distribution system to supply both heated and chilled water shall be designed to allow a dead band between changeover from one mode to the other of at least 15°F (8.3°C) outside air temperatures; be designed to and provided with controls that will allow operation in one mode for at least 4 hours before changing over to the other mode; and be provided with controls that allow heating and cooling supply temperatures at the changeover point to be no more than 30°F (16.7°C) apart.

503.4.3.3 Hydronic (water loop) heat pump systems. Hydronic heat pump systems shall comply with Sections 503.4.3.3.1 through 503.4.3.3.3.

503.4.3.3.1 Temperature dead band. Hydronic heat pumps connected to a common heat pump water loop with central devices for heat rejection and heat addition shall have controls that are capable of providing a heat pump water supply temperature dead band of at least 20°F (11.1°C) between initiation of heat rejection and heat addition by the central devices.

Exception: Where a system loop temperature optimization controller is installed and can determine the most efficient operating temperature based on real-time conditions of demand and capacity, dead bands of less than 20°F (11°C) shall be permitted.

503.4.3.3.2 Heat rejection. Heat rejection equipment shall comply with Sections 503.4.3.3.2.1 and 503.4.3.3.2.2.

Exception: Where it can be demonstrated that a heat pump system will be required to reject heat throughout the year.

503.4.3.3.2.1 Climate Zones 3 and 4. For Climate Zones 3 and 4 as indicated in Figure 301.1 and Table 301.1:

1. If a closed-circuit cooling tower is used directly in the heat pump loop, either an automatic valve shall be installed to bypass all but a minimal flow of water around the tower, or lower leakage positive closure dampers shall be provided.

2. If an open-circuit tower is used directly in the heat pump loop, an automatic valve shall be installed to bypass all heat pump water flow around the tower.

3. If an open- or closed-circuit cooling tower is used in conjunction with a separate heat exchanger to isolate the cooling tower from the heat pump loop, then heat loss shall be controlled by shutting down the circulation pump on the cooling tower loop.

503.4.3.3.2.2 Climate Zones 5 through 8. For climate Zones 5 through 8 as indicated in Figure 301.1 and Table 301.1, if an open- or closed-circuit cooling tower is used, then a separate heat exchanger shall be required to isolate the cooling tower from the heat pump loop, and heat loss shall be controlled by shutting down the circulation pump on the cooling tower loop and providing an automatic valve to stop the flow of fluid.

503.4.3.3.3 Two position valve. Each hydronic heat pump on the hydronic system having a total pump system power exceeding 10 horsepower (hp) (7.5kW) shall have a two-position valve.

503.4.3.4 Part load controls. Hydronic heating systems greater than or equal to 300,000 Btu/h (87,930

W) in design output capacity supplying heated or chilled water to comfort conditioning systems shall include controls that have the capability to:

1. Automatically reset the supply hot water temperatures using zone-return water temperature, building-return water temperature, or outside air temperature as an indicator of building heating or cooling demand. The temperature shall be capable of being reset by at least 25 percent of the design supply-to-return water temperature difference; or

2. Reduce system pump flow by at least 50 percent of design flow rate utilizing adjustable speed drive(s) on pump(s), or multiple-staged pumps where at least one-half of the total pump horsepower is capable of being automatically turned off or control valves designed to modulate or step down, and close, as a function of load, or other *approved* means.

503.4.3.5 Pump isolation. Chilled water plants including more than one chiller shall have the capability to reduce flow automatically through the chiller plant when a chiller is shut down. Chillers piped in series for the purpose of increased temperature differential shall be considered as one chiller. Boiler plants including more than one boiler shall have the capability to reduce flow automatically through the boiler plant when a boiler is shut down.

503.4.4 Heat rejection equipment fan speed control. Each fan powered by a motor of 7.5 hp (5.6 kW) or larger shall have the capability to operate that fan at two-thirds of full speed or less, and shall have controls that automatically change the fan speed to control the leaving fluid temperature or condensing temperature/pressure of the heat rejection device.

Exception: Factory-installed heat rejection devices within HVAC equipment tested and rated in accordance with Tables 503.2.3(6) and 503.2.3(7).

503.4.5 Requirements for complex mechanical systems serving multiple zones. Sections 503.4.5.1 through 503.4.5.3 shall apply to complex mechanical systems serving multiple zones. Supply air systems serving multiple zones shall be VAV systems which, during periods of occupancy, are designed and capable of being controlled to reduce primary air supply to each *zone* to one of the following before reheating, recooling or mixing takes place:

1. Thirty percent of the maximum supply air to each *zone*.

2. Three hundred cfm (142 L/s) or less where the maximum flow rate is less than 10 percent of the total fan system supply airflow rate.

3. The minimum ventilation requirements of Chapter 4 of the *International Mechanical Code*.

Exception: The following define when individual zones or when entire air distribution systems are exempted from the requirement for VAV control:

1. Zones where special pressurization relationships or cross-contamination requirements are such that VAV systems are impractical.

2. Zones or supply air systems where at least 75 percent of the energy for reheating or for providing warm air in mixing systems is provided from a site-recovered or site-solar energy source.

3. Zones where special humidity levels are required to satisfy process needs.

4. Zones with a peak supply air quantity of 300 cfm (142 L/s) or less and where the flow rate is less than 10 percent of the total fan system supply airflow rate.

5. Zones where the volume of air to be reheated, recooled or mixed is no greater than the volume of outside air required to meet the minimum ventilation requirements of Chapter 4 of the *International Mechanical Code.*

6. Zones or supply air systems with thermostatic and humidistatic controls capable of operating in sequence the supply of heating and cooling energy to the zone(s) and which are capable of preventing reheating, recooling, mixing or simultaneous supply of air that has been previously cooled, either mechanically or through the use of economizer systems, and air that has been previously mechanically heated.

503.4.5.1 Single duct variable air volume (VAV) systems, terminal devices. Single duct VAV systems shall use terminal devices capable of reducing the supply of primary supply air before reheating or recooling takes place.

503.4.5.2 Dual duct and mixing VAV systems, terminal devices. Systems that have one warm air duct and one cool air duct shall use terminal devices which are capable of reducing the flow from one duct to a minimum before mixing of air from the other duct takes place.

503.4.5.3 Single fan dual duct and mixing VAV systems, economizers. Individual dual duct or mixing heating and cooling systems with a single fan and with total capacities greater than 90,000 Btu/h [(26 375 W) 7.5 tons] shall not be equipped with air economizers.

503.4.5.4 Supply-air temperature reset controls. Multiple *zone* HVAC systems shall include controls that automatically reset the supply-air temperature in response to representative building loads, or to outdoor air temperature. The controls shall be capable of resetting the supply air temperature at least 25 percent of the difference between the design supply-air temperature and the design room air temperature.

Exceptions:

1. Systems that prevent reheating, recooling or mixing of heated and cooled supply air.

2. Seventy five percent of the energy for reheating is from site-recovered or site solar energy sources.

3. Zones with peak supply air quantities of 300 cfm (142 L/s) or less.

503.4.6 Heat recovery for service water heating. Condenser heat recovery shall be installed for heating or reheating of service hot water provided the facility operates 24 hours a day, the total installed heat capacity of water-cooled systems exceeds 6,000,000 Btu/hr of heat rejection, and the design service water heating load exceeds 1,000,000 Btu/h.

The required heat recovery system shall have the capacity to provide the smaller of:

1. Sixty percent of the peak heat rejection load at design conditions; or

2. The preheating required to raise the peak service hot water draw to 85°F (29°C).

Exceptions:

1. Facilities that employ condenser heat recovery for space heating or reheat purposes with a heat recovery design exceeding 30 percent of the peak water-cooled condenser load at design conditions.

2. Facilities that provide 60 percent of their service water heating from site solar or site recovered energy or from other sources.

503.4.7 Hot gas bypass limitation. Cooling systems shall not use hot gas bypass or other evaporator pressure control systems unless the system is designed with multiple steps of unloading or continuous capacity modulation. The capacity of the hot gas bypass shall be limited as indicated in Table 503.4.7.

Exception: Unitary packaged systems with nominal cooling capacities of 7.5 tons or less (approximately 90 kBtu/h or 26.4 kW).

TABLE 503.4.7
MAXIMUM HOT GAS BYPASS CAPACITY

RATED CAPACITY	MAXIMUM HOT GAS BYPASS CAPACITY (% of total capacity)
≤ 20 nominal tons (240 kBtu/h)	50%
> 20 nominal tons (240 kBtu/h)	25%

For SI: 1 Btu/h = 0.29 watts.

SECTION 504
SERVICE WATER HEATING
(Mandatory Requirements)

504.1 General. This section covers the minimum efficiency of, and controls for, service water-heating equipment and insulation of service hot water piping.

504.2 Service water-heating equipment performance efficiency. Water-heating equipment and hot water storage tanks shall meet the requirements of Table 504.2. The efficiency shall be verified through data furnished by the manufacturer or through certification under an *approved* certification program.

504.3 Temperature controls. Service water-heating equipment shall be provided with controls to allow a setpoint of 110°F (43°C) for equipment serving dwelling units and 90°F (32°C) for equipment serving other occupancies. The outlet temperature of lavatories in public facility rest rooms shall be limited to 110°F (43°C).

504.4 Heat traps. Water-heating equipment not supplied with integral heat traps and serving noncirculating systems shall be provided with heat traps on the supply and discharge piping associated with the equipment.

504.5 Pipe insulation. For automatic-circulating hot water systems, piping shall be insulated with 1 inch (25 mm) of insulation having a conductivity not exceeding 0.27 Btu per inch/h × ft^2 × °F (1.53 W per 25 mm/m^2 × K). The first 8 feet (2438 mm) of piping in noncirculating systems served by equipment without integral heat traps shall be insulated with 0.5 inch (12.7 mm) of material having a conductivity not exceeding 0.27 Btu per inch/h × ft^2 × °F (1.53 W per 25 mm/m^2 × K).

504.6 Hot water system controls. Automatic-circulating hot water system pumps or heat trace shall be arranged to be conveniently turned off automatically or manually when the hot water system is not in operation.

504.7 Pools and inground permanently installed spas (Mandatory). Pools and inground permanently installed spas shall comply with Sections 504.7.1 through 504.7.3.

504.7.1 Heaters. All heaters shall be equipped with a readily accessible on-off switch that is mounted outside of the heater to allow shutting off the heater without adjusting the thermostat setting. Gas-fired heaters shall not be equipped with constant burning pilot lights.

504.7.2 Time switches. Time switches or other control method that can automatically turn off and on heaters and pumps according to a preset schedule shall be installed on all heaters and pumps. Heaters, pumps and motors that have built-in timers shall be deemed in compliance with this requirement.

Exceptions:

1. Where public health standards require 24-hour pump operation.

2. Where pumps are required to operate solar- and waste-heat-recovery pool heating systems.

504.7.3 Covers. Heated pools and inground permanently installed spas shall be provided with a vapor-retardant cover.

Exception: Pools deriving over 70 percent of the energy for heating from site-recovered energy, such as a heat pump or solar energy source computed over an operating season.

TABLE 504.2
MINIMUM PERFORMANCE OF WATER-HEATING EQUIPMENT

EQUIPMENT TYPE	SIZE CATEGORY (input)	SUBCATEGORY OR RATING CONDITION	PERFORMANCE REQUIRED[a, b]	TEST PROCEDURE
Water heaters, Electric	≤ 12 kW	Resistance	0.97 - 0.00132V, EF	DOE 10 CFR Part 430
	> 12 kW	Resistance	1.73V + 155 SL, Btu/h	ANSI Z21.10.3
	≤ 24 amps and ≤ 250 volts	Heat pump	0.93 - 0.00132V, EF	DOE 10 CFR Part 430
Storage water heaters, Gas	≤ 75,000 Btu/h	≥ 20 gal	0.67 - 0.0019V, EF	DOE 10 CFR Part 430
	> 75,000 Btu/h and ≤ 155,000 Btu/h	< 4,000 Btu/h/gal	$80\% E_t$ $\left(Q/800 + 110\sqrt{V} \right)$ SL, Btu/h	ANSI Z21.10.3
	> 155,000 Btu/h	< 4,000 Btu/h/gal	$80\% E_t$ $\left(Q/800 + 110\sqrt{V} \right)$ SL, Btu/h	
Instantaneous water heaters, Gas	> 50,000 Btu/h and < 200,000 Btu/h[c]	≥ 4,000 (Btu/h)/gal and < 2 gal	0.62 - 0.0019V, EF	DOE 10 CFR Part 430
	≥ 200,000 Btu/h	≥ 4,000 Btu/h/gal and < 10 gal	$80\% E_t$	ANSI Z21.10.3
	≥ 200,000 Btu/h	≥ 4,000 Btu/h/gal and ≥ 10 gal	$80\% E_t$ $\left(Q/800 + 110\sqrt{V} \right)$ SL, Btu/h	
Storage water heaters, Oil	≤ 105,000 Btu/h	≥ 20 gal	0.59 - 0.0019V, EF	DOE 10 CFR Part 430
	> 105,000 Btu/h	< 4,000 Btu/h/gal	$78\% E_t$ $\left(Q/800 + 110\sqrt{V} \right)$ SL, Btu/h	ANSI Z21.10.3
Instantaneous water heaters, Oil	≤ 210,000 Btu/h	≥ 4,000 Btu/h/gal and < 2 gal	0.59 - 0.0019V, EF	DOE 10 CFR Part 430
	> 210,000 Btu/h	≥ 4,000 Btu/h/gal and < 10 gal	$80\% E_t$	ANSI Z21.10.3
	> 210,000 Btu/h	≥ 4,000 Btu/h/gal and ≥ 10 gal	$78\% E_t$ $\left(Q/800 + 110\sqrt{V} \right)$ SL, Btu/h	
Hot water supply boilers, Gas and Oil	≥ 300,000 Btu/h and <12,500,000 Btu/h	≥ 4,000 Btu/h/gal and < 10 gal	$80\% E_t$	ANSI Z21.10.3
Hot water supply boilers, Gas	≥ 300,000 Btu/h and <12,500,000 Btu/h	≥ 4,000 Btu/h/gal and ≥ 10 gal	$80\% E_t$ $\left(Q/800 + 110\sqrt{V} \right)$ SL, Btu/h	
Hot water supply boilers, Oil	> 300,000 Btu/h and <12,500,000 Btu/h	> 4,000 Btu/h/gal and > 10 gal	$78\% E_t$ $\left(Q/800 + 110\sqrt{V} \right)$ SL, Btu/h	
Pool heaters, Gas and Oil	All	—	$78\% E_t$	ASHRAE 146
Heat pump pool heaters	All	—	4.0 COP	AHRI 1160
Unfired storage tanks	All	—	Minimum insulation requirement R-12.5 (h · ft^2 · °F)/Btu	(none)

For SI: °C = [(°F) - 32]/1.8, 1 British thermal unit per hour = 0.2931 W, 1 gallon = 3.785 L, 1 British thermal unit per hour per gallon = 0.078 W/L.

a. Energy factor (EF) and thermal efficiency (E_t) are minimum requirements. In the EF equation, V is the rated volume in gallons.

b. Standby loss (SL) is the maximum Btu/h based on a nominal 70°F temperature difference between stored water and ambient requirements. In the SL equation, Q is the nameplate input rate in Btu/h. In the SL equation for electric water heaters, V is the rated volume in gallons. In the SL equation for oil and gas water heaters and boilers, V is the rated volume in gallons.

c. Instantaneous water heaters with input rates below 200,000 Btu/h must comply with these requirements if the water heater is designed to heat water to temperatures 180°F or higher.

SECTION 505
ELECTRICAL POWER AND LIGHTING SYSTEMS

505.1 General (Mandatory Requirements). This section covers lighting system controls, the connection of ballasts, the maximum lighting power for interior applications, and minimum acceptable lighting equipment for exterior applications.

Exception: Lighting within dwelling units where 75 percent or more of the permanently installed interior light fixtures are fitted with high-efficacy lamps.

505.2 Lighting controls (Mandatory Requirements). Lighting systems shall be provided with controls as required in Sections 505.2.1, 505.2.2, 505.2.3, and 505.2.4.

505.2.1 Interior lighting controls. Each area enclosed by walls or floor-to-ceiling partitions shall have at least one manual control for the lighting serving that area. The required controls shall be located within the area served by the controls or be a remote switch that identifies the lights served and indicates their status.

Exceptions:

1. Areas designated as security or emergency areas that must be continuously lighted.

2. Lighting in stairways or corridors that are elements of the means of egress.

505.2.2 Additional controls. Each area that is required to have a manual control shall have additional controls that meet the requirements of Sections 505.2.2.1 and 505.2.2.2.

505.2.2.1 Light reduction controls. Each area that is required to have a manual control shall also allow the occupant to reduce the connected lighting load in a reasonably uniform illumination pattern by at least 50 percent. Lighting reduction shall be achieved by one of the following or other *approved* method:

1. Controlling all lamps or luminaires;

2. Dual switching of alternate rows of luminaires, alternate luminaires or alternate lamps;

3. Switching the middle lamp luminaires independently of the outer lamps; or

4. Switching each luminaire or each lamp.

Exceptions:

1. Areas that have only one luminaire.

2. Areas that are controlled by an occupant-sensing device.

3. Corridors, storerooms, restrooms or public lobbies.

4. *Sleeping unit* (see Section 505.2.3).

5. Spaces that use less than 0.6 watts per square foot (6.5 W/m²).

505.2.2.2 Automatic Lighting Shutoff.

505.2.2.2.1 Building automatic lighting controls. Buildings larger than 5,000 square feet (465 m²) shall be equipped with an automatic control device to shut off lighting. This automatic control requirement shall be achieved by one or a combination of the following:

1. A scheduled basis, using time-of-day, with an independent program schedule that controls the interior lighting in areas that do not exceed 25,000 square feet (2,323 m²) and are not more than one floor;

2. A signal from another control or alarm system that indicates the area is unoccupied; or

3. An occupancy sensor that automatically turns lighting off within 30 minutes of an occupant leaving a space.

> **Exception:** The following shall not require an automatic control device:
>
> 1. Sleeping unit (see Section 505.2.3).
>
> 2. Lighting in spaces where patient care is directly provided.
>
> 3. Spaces where an automatic shutoff would endanger occupant safety or security.
>
> 4. Spaces with daylight control zones that have automatic daylighting controls using either continuous or stepped control systems.
>
> 5. Lighting intended for 24-hour operation.
>
> 6. Tenant spaces less than 2,000 square feet complying with Section 505.2.2.2.2.

505.2.2.2.2 Occupancy sensors. All buildings shall have occupancy sensors in all of the following spaces:

1. Classrooms;

2. Conference/meeting rooms;

3. Employee lunch and break rooms;

4. Private offices;

5. Storage rooms over 100 square feet; and

6. Computer rooms over 100 square feet.

505.2.2.2.3 Occupant override. Where an automatic time switch control device is installed to comply with Section 505.2.2.2.1, Item 1, it shall incorporate an override switching device that:

1. Is readily *accessible*.

2. Is located so that a person using the device can see the lights or the area controlled by that switch, or so that the area being lit is annunciated.

3. Is manually operated.

4. Allows the lighting to remain on for no more than 2 hours when an override is initiated.

5. Controls an area not exceeding 5,000 square feet (465 m²).

Exceptions:

1. In malls and arcades, auditoriums, single-tenant retail spaces, industrial facilities and arenas, where captive-key override is utilized, override time shall be permitted to exceed 2 hours.

2. In malls and arcades, auditoriums, single-tenant retail spaces, industrial facilities and arenas, the area controlled shall not exceed 20,000 square feet (1860 m²).

505.2.2.2.4 Holiday scheduling. If an automatic time switch control device is installed in accordance with Section 505.2.2.2, Item 1, it shall incorporate an automatic holiday scheduling feature that turns off all loads for at least 24 hours, then resumes the normally scheduled operation.

Exception: Retail stores and associated malls, restaurants, grocery stores, places of religious worship and theaters.

505.2.3 Sleeping unit controls. *Sleeping units* in hotels, motels, boarding houses or similar buildings shall have at least one master switch at the main entry door that controls all permanently wired luminaires and switched receptacles, except those in the bathroom(s). Suites shall have a control meeting these requirements at the entry to each room or at the primary entry to the suite.

505.2.4 Exterior lighting controls. Lighting not designated for dusk-to-dawn operation shall be controlled by either a combination of a photosensor and a time switch, or an astronomical time switch. Lighting designated for dusk-to-dawn operation shall be controlled by an astronomical time switch or photosensor. All time switches shall be capable of retaining programming and the time setting during loss of power for a period of at least 10 hours.

505.2.5 Additional miscellaneous lighting control.

505.2.5.1 Separate lighting controls. Separate lighting controls shall be required for:

1. Display/Accent Lighting;

2. Case Lighting;

3. Nonvisual Lighting—lighting for nonvisual applications, such as plant growth and food warming; and

4. Demonstration Lighting—lighting equipment that is for sale or for demonstrations in lighting education.

505.2.5.2 Hotel and motel guest room lighting. Hotel and motel guest rooms and guest suites shall have a master control device at the main room entry that controls all permanently installed luminaires and switched receptacles.

505.2.5.3 Task lighting. Supplemental task lighting, including permanently installed undershelf or undercabinet lighting, shall have a control device integral to the luminaires or be controlled by a wall-mounted control device provided the control device is readily accessible and located so that the occupant can see the controlled lighting.

505.2.6 Lighting controls functional performance testing requirements. For lighting controls which include daylight or occupant sensing automatic controls, automatic shut-off controls, occupancy sensors, or automatic time switches, the lighting control shall be tested to ensure that control devices, components, equipment, and systems are calibrated, adjusted and operate in accordance with approved plans and specifications. Sequences of operation shall be functionally tested to ensure they operate in accordance with approved plans and specifications.

505.3 Tandem wiring (Mandatory Requirements). The following luminaires located within the same area shall be tandem wired:

1. Fluorescent luminaires equipped with one, three or odd-numbered lamp configurations, that are recess mounted within 10 feet (3048 mm) center-to-center of each other.

2. Fluorescent luminaires equipped with one, three or any odd-numbered lamp configuration, that are pendant- or surface-mounted within 1 foot (305 mm) edge-to-edge of each other.

Exceptions:

1. Where electronic high-frequency ballasts are used.

2. Luminaires on emergency circuits.

3. Luminaires with no available pair in the same area.

505.4 Exit signs (Mandatory Requirements). Internally illuminated exit signs shall not exceed 5 watts per side.

505.5 Interior lighting power requirements (Prescriptive). A building complies with this section if its total connected lighting power calculated under Section 505.5.1 is no greater than the interior lighting power calculated under Section 505.5.2.

505.5.1 Total connected interior lighting power. The total connected interior lighting power (watts) shall be the sum of the watts of all interior lighting equipment as determined in accordance with Sections 505.5.1.1 through 505.5.1.4.

Exceptions:

1. The connected power associated with the following lighting equipment is not included in calculating total connected lighting power.

1.1. Professional sports arena playing field lighting.

1.2. *Sleeping unit* lighting in hotels, motels, boarding houses or similar buildings.

1.3. Emergency lighting automatically off during normal building operation.

1.4. Lighting in spaces designed for use by occupants with special lighting needs including the visually impaired visual impairment and other medical and age-related issues.

 1.5. Lighting in interior spaces that have been designated as a registered interior historic landmark.

 1.6. Casino gaming areas.

2. Lighting equipment used for the following shall be exempt provided that it is in addition to general lighting and is controlled by an independent control device:

 2.1. Task lighting for medical and dental purposes.

 2.2. Display lighting for exhibits in galleries, museums and monuments.

3. Lighting for theatrical purposes, including performance, stage, film production and video production.

4. Lighting for photographic processes.

5. Lighting integral to equipment or instrumentation and is installed by the manufacturer.

6. Task lighting for plant growth or maintenance.

7. Advertising signage or directional signage.

8. In restaurant buildings and areas, lighting for food warming or integral to food preparation equipment.

9. Lighting equipment that is for sale.

10. Lighting demonstration equipment in lighting education facilities.

11. Lighting *approved* because of safety or emergency considerations, inclusive of exit lights.

12. Lighting integral to both open and glass enclosed refrigerator and freezer cases.

13. Lighting in retail display windows, provided the display area is enclosed by ceiling-height partitions.

14. Furniture mounted supplemental task lighting that is controlled by automatic shutoff.

505.5.1.1 Screw lamp holders. The wattage shall be the maximum *labeled* wattage of the luminaire.

505.5.1.2 Low-voltage lighting. The wattage shall be the specified wattage of the transformer supplying the system.

505.5.1.3 Other luminaires. The wattage of all other lighting equipment shall be the wattage of the lighting equipment verified through data furnished by the manufacturer or other *approved* sources.

505.5.1.4 Line-voltage lighting track and plug-in busway. The wattage shall be:

1. The specified wattage of the luminaires included in the system with a minimum of 30 W/lin ft. (98 W/lin. m);

2. The wattage limit of the system's circuit breaker; or

3. The wattage limit of other permanent current limiting device(s) on the system.

505.5.2 Interior lighting power. Interior lighting power allowance shall be calculated using one of the following two methods:

1. **Building area method:**

 (a) Determine the appropriate building area type from Table 505.5.2 and the allowed lighting power density (watts per unit area) from the building area method column. For building area types not listed, selection of a reasonably equivalent type shall be permitted.

 (b) Determine the gross lighted floor area (square feet) of the building area type.

 (c) Multiply the gross lighted floor areas of the building area type(s) times the lighting power density.

 (d) The interior lighting power allowance for the building is the sum of the lighting power allowances of all building area types. Trade-offs among building area types are permitted provided that the total installed interior lighting power does not exceed the interior lighting power allowance.

2. **Space-by-space method:**

 (a) Determine the appropriate building type from Table 505.5.2. For building types not listed, selection of a reasonably equivalent type shall be permitted.

 (b) For each space enclosed by partitions 80 percent or greater than ceiling height, determine the gross interior floor area by measuring to the center of the partition wall. Include the floor area of balconies or other projections. Retail spaces do not have to comply with the 80 percent partition height requirements.

 (c) Determine the interior lighting power allowance by using the columns designated space-by-space method in Table 505.5.2. Multiply the floor area(s) of the space(s) times the allowed lighting power density for the space type that most closely represents the proposed use of the space(s). The product is the lighting power allowance for the space(s). For space types not listed, selection of a reasonably equivalent category shall be permitted.

 (d) The interior lighting power allowance is the sum of lighting power allowances of all spaces. Trade-offs among spaces are permitted provided that the total installed interior lighting power does not exceed the interior lighting power allowance.

TABLE 505.5.2
INTERIOR LIGHTING POWER ALLOWANCES

Building Area Type[a]	Whole Building	Space by Space (W/ft^2)
LIGHTING POWER DENSITY		
Common Space Types		
Active Storage		0.63
Atrium - First Three Floors		0.63
Atrium - Each Additional Floor		0.16
Automotive Facility		0.91
Bank/Office, Bank Activity Area		1.38
Classroom/Lecture/Training		1.25
Conference/Meeting/Multipurpose		1.29
Corridor/Transition		0.65
Education Laboratory		1.28
Electrical/Mechanical		0.95
Food Preparation		0.99
Lobby		0.60
Locker Room		0.78
Medical/Industrial Research Laboratory		1.62
Parking Garage - Garage Area		0.21
Restroom		0.84
Stairway		0.69
Convention Center	1.05	
Exhibit Space		1.58
Audience/Seating Area		0.80
Court House	1.07	
Audience/Seating Area		0.80
Courtroom		1.91
Confinement Cells		1.10
Judges Chambers		1.17
Dressing/Locker/Fitting Room		0.78
Dining: Bar/Lounge/Leisure	1.01	
Lounge/Leisure Dining		1.40
Dining: Cafeteria/Fast Food	0.93	
Dining: Family	0.94	
Dining		0.99
Dormitory	0.58	
Living Quarters		0.32
Bedroom		0.50
Study Hall		1.30
Exercise Center	0.89	
Dressing/Locker/Fitting Room		0.78
Audience/Seating Area		0.30
Exercise Area/Gymnasium		0.72
Gymnasium	0.70	
Dressing/Locker/Fitting Room		0.78
Audience/Seating Area		0.40
Playing Area		1.35
Exercise Area		0.72
Healthcare Clinic/Hospital	1.06	
Corridors w/patient waiting, exam		0.94
Exam/Treatment		1.66
Emergency		2.35
Public & Staff Lounge		0.79
Hospital/Medical Supplies		1.27

continued

TABLE 505.5.2—continued
INTERIOR LIGHTING POWER ALLOWANCES

Building Area Type[a]	Whole Building	Space by Space (W/ft^2)
LIGHTING POWER DENSITY		
Common Space Types		
Hospital - Nursery		0.60
Nurse Station		0.87
Physical Therapy		0.91
Patient Room		0.62
Pharmacy		1.14
Hospital/Radiology		1.34
Operating Room		1.89
Recovery		1.15
Active Storage		0.63
Laundry – Washing		0.60
Hotel	1.01	
Dining Area		0.85
Guest quarters		1.11
Reception/Waiting		2.30
Lobby		1.05
Library	0.96	
Library - Audio Visual		0.60
Stacks		1.42
Card File & Cataloguing		0.72
Reading Area		0.93
Manufacturing Facility	0.98	
Low Bay (< 25 ft Floor to Ceiling)		1.19
High Bay (> 25 ft Floor to Ceiling)		1.34
Detailed Manufacturing		1.29
Equipment Room		0.95
Corridor/Transition		0.41
Motel	1.05	
Dining Area		1.05
Living Quarters		0.75
Reception/Waiting		1.90
Motion Picture Theater	0.86	
Audience/Seating Area		0.53
Lobby		1.13
Multi-Family	0.53	
Museum	1.05	
Active Storage		0.63
General Exhibition		1.05
Restoration		1.02
Office	0.89	
Enclosed		1.11
Open Plan		0.98
Parking Garage	0.22	
Penitentiary	0.94	
Performing Arts Theater	1.35	
Audience/Seating Area		2.30
Lobby		2.34
Dressing/Locker/Fitting Room		1.14
Police/Fire Station	0.94	
Fire Station Engine Room		0.56
Sleeping Quarters		0.25
Post Office/SF	0.84	
Sorting Area		0.94
Lobby		1.00

continued

TABLE 505.5.2—continued
INTERIOR LIGHTING POWER ALLOWANCES

| Building Area Type[a] | LIGHTING POWER DENSITY | |
| | Whole Building | Space by Space |
	(W/ft²)	
Common Space Types		
Religious Buildings	1.14	
Lobby/Fellowship Hall		0.64
Worship/Pulpit/Choir		1.95
Retail[c]	1.41	
Department Store Sales Area		1.30
Dressing/Fitting Room		0.96
Fine Merchandise Sales Area		2.60
Mall Concourse		1.06
Mass Merchandising Sales Area		1.30
Personal Services Sales Area		1.30
Specialty Store Sales Area		1.60
Supermarket Sales Area		1.30
School/University	0.98	
Classroom		1.30
Audience		0.70
Dining Area		1.00
Office		1.00
Corridor		0.50
Storage		0.50
Laboratory		1.28
Sports Arena	0.71	
Ring Sports Arena		2.68
Court Sports Arena		1.80
Town Hall	0.89	
Transportation	0.76	
Dining Area		1.90
Baggage Area		0.76
Airport - Concourse		0.39
Terminal - Ticket Counter		1.12
Reception/Waiting		0.50
Warehouse	0.56	
Fine Material		0.95
Medium Bulky Material		0.63
Workshop	1.59	

For SI: 1 foot = 304.8 mm, 1 watt per square foot = W/0.0929 m².

a. In cases where both a general building area type and a more specific building area type are listed, the more specific building area type shall apply.

b. The lighting power densities contained in this table include allowances for video-display terminals, decorative lighting and display lighting Additional lighting power is not allowed for these uses. Task lighting is not included in these connected LPD limits.

c. Where lighting equipment is specified to be installed to highlight specific merchandise in addition to lighting equipment specified for general lighting and is switched or dimmed on circuits different from the circuits for general lighting, the smaller of the actual wattage of the lighting equipment installed specifically for merchandise, or additional lighting power as determined below shall be added to the interior lighting power determined in accordance with this line item.

Calculate the additional lighting power as follows:

Additional Interior Lighting Power Allowance = (Retail Area 1 × 0.4W/ft²) + (Retail Area 2 × 0.6 W/ft²) + (Retail Area 3 × 0.9 W/ft²) + (Retail Area 4 × 1.5 W/ft²).

where:

Retail Area 1 = The floor area for all products not listed in Retail Area 2, 3 or 4.

Retail Area 2 = The floor area used for the sale of vehicles, sporting goods and small electronics.

Retail Area 3 = The floor area used for the sale of furniture, clothing, cosmetics and artwork.

Retail Area 4 = The floor area used for the sale of jewelry, crystal and china.

Exception: Other merchandise categories are permitted to be included in Retail Areas 2 through 4 above, provided that justification documenting the need for additional lighting power based on visual inspection, contrast, or other critical display is *approved* by the authority having jurisdiction.

505.6 Exterior lighting. (Mandatory Requirements). When the power for exterior lighting is supplied through the energy service to the building, all exterior lighting, other than low-voltage landscape lighting, shall comply with Sections 505.6.1 and 505.6.2.

Exception: Where *approved* because of historical, safety, signage or emergency considerations.

505.6.1 Exterior building grounds lighting. All exterior building grounds luminaires that operate at greater than 100 watts shall contain lamps having a minimum efficacy of 60 lumens per watt unless the luminaire is controlled by a motion sensor or qualifies for one of the exceptions under Section 505.6.2.

505.6.2 Exterior building lighting power. The total exterior lighting power allowance for all exterior building applications is the sum of the base site allowance plus the individual allowances for areas that are to be illuminated and are permitted in Table 505.6.2(2) for the applicable lighting *zone*. Tradeoffs are allowed only among exterior lighting applications listed in Table 505.6.2(2), Tradable Surfaces section. The lighting zone for the building exterior is determined from Table 505.6.2(1) unless otherwise specified by the local jurisdiction. Exterior lighting for all applications (except those included in the exceptions to Section 505.6.2) shall comply with the requirements of Section 505.6.1.

Exceptions: Lighting used for the following exterior applications is exempt when equipped with a control device independent of the control of the nonexempt lighting:

1. Specialized signal, directional and marker lighting associated with transportation;

2. Advertising signage or directional signage;

3. Integral to equipment or instrumentation and is installed by its manufacturer;

4. Theatrical purposes, including performance, stage, film production and video production;

5. Athletic playing areas;

6. Temporary lighting;

7. Industrial production, material handling, transportation sites and associated storage areas;

8. Theme elements in theme/amusement parks; and

9. Used to highlight features of public monuments and registered historic landmark structures or buildings.

TABLE 505.6.2(1)
EXTERIOR LIGHTING ZONES

LIGHTING ZONE	DESCRIPTION
1	Developed areas of national parks, state parks, forest land, and rural areas
2	Areas predominantly consisting of residential zoning, neighborhood business districts, light industrial with limited nighttime use and residential mixed use areas
3	All other areas
4	High-activity commercial districts in major metropolitan areas as designated by the local land use planning authority

TABLE 505.6.2(2)
INDIVIDUAL LIGHTING POWER ALLOWANCES FOR BUILDING EXTERIORS

		Zone 1	Zone 2	Zone 3	Zone 4
Base Site Allowance (Base allowance may be used in tradable or nontradable surfaces.)		500 W	600 W	750 W	1300 W
Tradable Surfaces (Lighting power densities for uncovered parking areas, building grounds, building entrances and exits, canopies and overhangs and outdoor sales areas may be traded.)	**Uncovered Parking Areas**				
	Parking areas and drives	0.04 W/ft^2	0.06 W/ft^2	0.10 W/ft^2	0.13 W/ft^2
	Building Grounds				
	Walkways less than 10 feet wide	0.7 W/linear foot	0.7 W/linear foot	0.8 W/linear foot	1.0 W/linear foot
	Walkways 10 feet wide or greater, plaza areas, special feature areas	0.14 W/ft^2	0.14 W/ft^2	0.16 W/ft^2	0.2 W/ft^2
	Stairways	0.75 W/ft^2	1.0 W/ft^2	1.0 W/ft^2	1.0 W/ft^2
	Pedestrian tunnels	0.15 W/ft^2	0.15 W/ft^2	0.2 W/ft^2	0.3 W/ft^2
	Building Entrances and Exits				
	Main entries	20 W/linear foot of door width	20 W/linear foot of door width	30 W/linear foot of door width	30 W/linear foot of door width
	Other doors	20 W/linear foot of door width	20 W/linear foot of door width	20 W/linear foot of door width	20 W/linear foot of door width
	Entry canopies	0.25 W/ft^2	0.25 W/ft^2	0.4 W/ft^2	0.4 W/ft^2
	Sales Canopies				
	Free-standing and attached	0.6 W/ft^2	0.6 W/ft^2	0.8 W/ft^2	1.0 W/ft^2
	Outdoor Sales				
	Open areas (including vehicle sales lots)	0.25 W/ft^2	0.25 W/ft^2	0.5 W/ft^2	0.7 W/ft^2
	Street frontage for vehicle sales lots in addition to "open area" allowance	No allowance	10 W/linear foot	10 W/linear foot	30 W/linear foot
Nontradable Surfaces (Lighting power density calculations for the following applications can be used only for the specific application and cannot be traded between surfaces or with other exterior lighting. The following allowances are in addition to any allowance otherwise permitted in the "Tradable Surfaces" section of this table.)	Building facades	No allowance	0.1 W/ft^2 for each illuminated wall or surface or 2.5 W/linear foot for each illuminated wall or surface length	0.15 W/ft^2 for each illuminated wall or surface or 3.75 W/linear foot for each illuminated wall or surface length	0.2 W/ft^2 for each illuminated wall or surface or 5.0 W/linear foot for each illuminated wall or surface length
	Automated teller machines and night depositories	270 W per location plus 90 W per additional ATM per location	270 W per location plus 90 W per additional ATM per location	270 W per location plus 90 W per additional ATM per location	270 W per location plus 90 W per additional ATM per location
	Entrances and gatehouse inspection stations at guarded facilities	0.75 W/ft^2 of covered and uncovered area	0.75 W/ft^2 of covered and uncovered area	0.75 W/ft^2 of covered and uncovered area	0.75 W/ft^2 of covered and uncovered area
	Loading areas for law enforcement, fire, ambulance and other emergency service vehicles	0.5 W/ft^2 of covered and uncovered area	0.5 W/ft^2 of covered and uncovered area	0.5 W/ft^2 of covered and uncovered area	0.5 W/ft^2 of covered and uncovered area
	Drive-up windows/doors	400 W per drive-through	400 W per drive-through	400 W per drive-through	400 W per drive-through
	Parking near 24-hour retail entrances	800 W per main entry	800 W per main entry	800 W per main entry	800 W per main entry

For SI: 1 foot = 304.8 mm, 1 watt per square foot = W/0.0929 m^2.

TABLE 506.2.1(1)
UNITARY AIR CONDITIONERS AND CONDENSING UNITS, ELECTRICALLY OPERATED, EFFICIENCY REQUIREMENTS

EQUIPMENT TYPE	SIZE CATEGORY[c]	SUBCATEGORY OR RATING CONDITION	REQUIRED EFFICIENCY[a]
Air conditioners, Air cooled	< 65,000 Btu/hd	Split system	15.0 SEER 12.5 EER
		Single package	15.0 SEER 12.0 EER
	≥ 65,000 Btu/h and < 135,000 Btu/h	Split system and single package	12.0 EER[b] 12.4 IPLV[b]
	≥ 135,000 Btu/h and < 240,000 Btu/h	Split system and single package	12.0 EER[b] 12.4 IPLV[b]
	≥ 240,000 Btu/h and < 760,000 Btu/h	Split system and single package	10.8 EER[b] 11.0 IPLV[b]
	≥ 760,000 Btu/h		10.2 EER[b] 11.0 IPLV[b]
Air conditioners, Water and evaporatively cooled		Split system and single package	14.0 EER

For SI: 1 British thermal unit per hour = 0.2931 W.

a. IPLVs are only applicable to equipment with capacity modulation.

b. Deduct 0.2 from the required EERs and IPLVs for units with a heating section other than electric resistance heat.

c. Size category is based on nominal equipment sizes, e.g. < 65,000 Btu/h indicates ≤ 5 tons, < 135,000 Btu/h indicates ≤ 10 tons, < 240,000 Btu/h indicates < 20 tons, ≥ 760,000 Btu/h indicates > 60 tons.

TABLE 506.2.1(2)
UNITARY AND APPLIED HEAT PUMPS, ELECTRICALLY OPERATED, EFFICIENCY REQUIREMENTS

EQUIPMENT TYPE	SIZE CATEGORY[c]	SUBCATEGORY OR RATING CONDITION	REQUIRED EFFICIENCY[a]
Air cooled (Cooling mode)	< 65,000 Btu/h[d]	Split system	15.0 SEER 12.5 EER
		Single package	15.0 SEER 12.0 EER
	≥ 65,000 Btu/h and < 135,000 Btu/h	Split system and single package	12.4 EER[b] 11.9 IPLV[b]
	≥ 135,000 Btu/h and < 240,000 Btu/h	Split system and single package	12.4 EER[b] 11.9 IPLV[b]
	≥ 240,000 Btu/h	Split system and single package	12.4 EER[b] 10.9 IPLV[b]
Water source (Cooling mode)	< 135,000 Btu/h	85°F entering water	14.0 EER
Air cooled (Heating mode)	< 65,000 Btu/h[d] (Cooling capacity)	Split system	9.0 HSPF
		Single package	8.5 HSPF
	≥ 65,000 Btu/h and < 135,000 Btu/h (Cooling capacity)	47°F db/43°F wb outdoor air	3.4 COP
		77°F db/15°F wb outdoor air	2.4 COP
	≥ 135,000 Btu/h (Cooling capacity)	47°F db/43°F wb outdoor air	3.2 COP
		77°F db/15°F wb outdoor air	2.1 COP
Water source (Heating mode)	< 135,000 Btu/h (Cooling capacity)	70°F entering water	4.6 COP

For SI: °C = [(°F) - 32] / 1.8, 1 British thermal unit per hour = 0.2931 W.

db = dry-bulb temperature, °F; wb = wet-bulb temperature, °F

a. IPLVs and Part load rating conditions are only applicable to equipment with capacity modulation.

b. Deduct 0.2 from the required EERs and IPLVs for units with a heating section other than electric resistance heat.

c. Size category is based on nominal equipment sizes, e.g. < 65,000 Btu/h indicates ≤ 5 tons, < 135,000 Btu/h indicates ≤ 10 tons, < 240,000 Btu/h indicates < 20 tons, ≥ 760,000 Btu/h indicates > 60 tons.

TABLE 506.2.1(3)
PACKAGED TERMINAL AIR CONDITIONERS AND PACKAGED TERMINALHEAT PUMPS

EQUIPMENT TYPE	SIZE CATEGORY	REQUIRED EFFICIENCY[b]
Air conditioners & Heat Pumps (Cooling Mode)	< 7,000 Btu/h	11.9 EER
	7,000 Btu/h and < 10,000 Btu/h	11.3 EER
	10,000 Btu/h and < 13,000 Btu/h	10.7 EER
	> 13,000 Btu/h	9.5 EER

a. Replacement units must be factory labeled as follows: "MANUFACTURED FOR REPLACEMENT APPLICATIONS ONLY: NOT TO BE INSTALLED IN NEW CONSTRUCTION PROJECTS." Replacement efficiencies apply only to units with existing sleeves less than 16 inches (406 mm) high and less than 42 inches (1067 mm) wide.

TABLE 506.2.1(4)
WARM AIR FURNACES AND COMBINATION WARM AIR FURNACES/AIR-CONDITIONING UNITS,
WARM AIR DUCT FURNACES AND UNIT HEATERS, EFFICIENCY REQUIREMENTS

EQUIPMENT TYPE	SIZE CATEGORY (INPUT)	SUBCATEGORY OR RATING CONDITION	REQUIRED EFFICIENCY
Warm air furnaces, gas fired	< 225,000 Btu/h	—	92 AFUE or 92 E_t
	≥ 225,000 Btu/h	Maximum capacity	90% E_c note 1
Warm air furnaces, oil fired	< 225,000 Btu/h	—	85 AFUE or 85 E_t
	≥ 225,000 Btu/h	Maximum capacity	85% E_t, Note 1
Warm air duct furnaces, gas fired	All capacities	Maximum capacity	90% E_c
Warm air unit heaters, gas fired	All capacities	Maximum capacity	90% E_c
Warm air unit heaters, oil fired	All capacities	Maximum capacity	90% E_c

For SI: 1 British thermal unit per hour = 0.2931 W.

1. Units must also include an IID (intermittent ignition device), have jackets not exceeding 0.75 percent of the input rating, and have either power venting or a flue damper. A vent damper is an acceptable alternative to a flue damper for those furnaces where combustion air is drawn from the conditioned space. Where there two ratings units not covered by the National Appliance Energy Conservation Act of 1987 (NAECA) (3-phase power or cooling capacity greater than or equal to 65,000 Btu/h [19 kW]) shall comply with either rating.

E_t = Thermal efficiency.

E_c = Combustion efficiency (100% less flue losses).

Efficient furnace fan: All fossil fuel furnaces shall have a furnace electricity ratio not greater than 0.02 and shall include a manufacturer's designation of the furnace electricity ratio.

TABLE 506.2.1(5)
BOILER, EFFICIENCY REQUIREMENTS

EQUIPMENT TYPE	SIZE CATEGORY	TEST PROCEDURE	REQUIRED EFFICIENCY
Gas Hot Water	< 300,000 Btu/h	DOE 10 CFR Part 430	90% E_t
	> 300,000 Btu/h and > 2.5 m Btu/h	DOE 10 CFR Part 431	89% E_t
Gas Steam	< 300,000 Btu/h	DOE 10 CFR Part 430	89% E_t
	> 300,000 Btu/h	DOE 10 CFR Part 431	89% E_t
Oil	< 300,000 Btu/h	DOE 10 CFR Part 430	90% E_t
	> 300,000 Btu/h	DOE 10 CFR Part 431	89% E_t

E_t = thermal efficiency

* Systems must be designed with lower operating hot water temperatures (<150°F) and use hot water reset to take advantage of the much higher efficiencies of condensing boilers.

TABLE 506.2.1(6)
CHILLERS - EFFICIENCY REQUIREMENTS

EQUIPMENT TYPE	SIZE CATEGORY	REQUIRED EFFICIENCY-CHILLERS		OPTIONAL COMPLIANCE PATH - REQUIRED EFFICIENCY - CHILLERS WITH VSD	
		IPLV (KW /TON)	Full Load (KW /TON)	Full Load (KW /TON)	IPLV (KW /TON)
Air Cooled w/ Condenser	All	1.2	1.0	N/A	N/A
Air Cooled w/o Condenser	All	1.08	1.08	N/A	N/A
Water Cooled, Reciprocating	All	0.840	0.630	N/A	N/A
Water Cooled, Rotary Screw and Scroll	< 90 tons	0.780	0.600	N/A	N/A
	90 tons and < 150 tons	0.730	0.550	N/A	N/A
	150 tons and < 300 tons	0.610	0.510	N/A	N/A
	> 300 tons	0.600	0.490	N/A	N/A
Water Cooled, Centrifugal	< 150 tons	0.610	0.620	0.630	0.400
	150 tons and < 300 tons	0.590	0.560	0.600	0.400
	300 tons and < 600 tons	0.570	0.510	0.580	0.400
	> 600 tons	0.550	0.510	0.550	0.400

a. Compliance with full load efficiency numbers and IPLV numbers are both required.
b. Only Chillers with Variable Speed Drives(VSD) may use the optional compliance path here for chiller efficiency.

TABLE 506.2.1(7)
ABSORPTION CHILLERS - EFFICIENCY REQUIREMENTS

EQUIPMENT TYPE	REQUIRED EFFICIENCY FULL LOAD COP (IPLV)
Air Cooled, Single Effect	0.60, but only allowed in heat recovery applications
Water Cooled, Single Effect	0.70, but only allowed in heat recovery applications
Double Effect - Direct Fired	1.0 (1.05)
Double Effect - Indirect Fired	1.20

505.6.3 Shielding of exterior building lighting fixtures.
Only fully shielded fixtures shall be permitted unless a lighting plan is submitted showing that the use of alternative fixtures would provide greater energy efficiency than any comparable lighting plan using fully shielded fixtures.

Exceptions:

1. Luminaires with an output of 150 Watts or less, or the equivalent light output.

2. Luminaires intended to illuminate the façade of buildings or to illuminate other objects including flagpoles, landscape and water features, statuary and works of art.

3. Luminaires for historic lighting on the premises of an historic building as defined in the *North Carolina State Building Codes* or within a designated historic district.

4. Outdoor sports facility lighting of the participant sport area.

5. Emergency exit discharge lighting.

6. Low voltage landscape lighting.

7. Sign illumination.

8. Festoon lighting as defined in the NFPA 70.

9. Temporary lighting for emergency, repair, construction, special events or similar activities.

505.7 Electrical energy consumption. (Mandatory Requirements). In buildings having individual dwelling units, provisions shall be made to determine the electrical energy consumed by each tenant by separately metering individual dwelling units.

SECTION 506
ADDITIONAL PRESCRIPTIVE
COMPLIANCE REQUIREMENTS

506.1 Requirements. Commercial buildings are required to comply with one of the following sections:

1. 506.2.1 More Efficient Mechanical Equipment;

2. 506.2.2 Reduced Lighting Power Density;

3. 506.2.3 Energy Recovery Ventilation Systems;

4. 506.2.4 Higher Efficiency Service Water Heating;

5. 506.2.5 On-Site Supply of Renewable Energy; or

6. 506.2.6 Automatic Daylighting Control System.

At the time of plan submittal, the building jurisdiction shall be provided, by the submittal authority, documentation designating the intent to comply with Section 506.2.1, 506.2.2, 506.2.3, 506.2.4, 506.2.5, or 506.2.6.

506.2.1 Efficient mechanical equipment. This mechanical alternative compliance option is intended to allow the builder to meet the requirements of Section 506 by choosing to install efficient mechanical equipment.

Mechanical equipment choices that fulfill requirements for Section 506.2.1 shall comply with the following in addition to the requirements in Section 503:

1. Package unitary equipment shall meet the minimum efficiency requirements in Tables 506.2.1(1) and 506.2.1(2);

2. Package terminal air conditioners and heat pumps shall meet the minimum efficiency requirements in Table 506.2.1(3);

3. Warm air furnaces and combination warm air furnaces/air conditioning units shall meet the minimum efficiency requirements in Table 506.2.1(4);

4. Boilers shall meet the minimum efficiency requirements in Table 506.2.1(5);

5. Electric chillers shall meet the energy efficiency requirements in Table 506.2.1(6); and

6. Absorption chillers shall meet the minimum efficiency requirements in Table 506.2.1(7).

506.2.2 Reduced lighting power density. Whole Building Lighting Power Density (watts/SF) must comply with values 10 percent lower than those in Table 505.5.2. Compliance with this section is in addition to meeting the requirements of Section 505.

506.2.3 Energy recovery ventilation systems. Buildings using 500 cfm or more outdoor air shall have heat or energy recovery ventilation systems for at least 80 percent of ventilation air. The recovery system shall provide a change in the enthalpy of the outdoor air supply of 50 percent or more of the difference between the outdoor air and return air at design conditions. Provision shall be made to bypass or control the energy recovery system to permit cooling with outdoor air where cooling with outdoor air is required.

506.2.4 High efficiency service water heating. Buildings must be of the following types to use this compliance method:

1. Hotels or motels;

2. Hospitals;

3. Restaurants or buildings containing food preparation areas;

4. Buildings with residential occupancies;

5. Buildings with laundry facilities or other high process service water heating needs; or

6. Buildings showing a service hot water load of 8 percent or more of total building energy loads as shown with an energy analysis as described in Section 507.

The building service water heating system shall have one of the following:

1. Instantaneous fuel-fired water heating systems for all fuel-fired water heating systems;

2. Electric heat pump water heating systems;

3. Water heating provided by geothermal heat pumps; or

4. Solar water heating systems sized to provide at least 40 percent of hot water requirements.

506.2.5 On-site supply of renewable energy. The building or surrounding property shall incorporate an on-site renewable energy system that supplies 3 percent or more of total building energy loads. On-site power generation using nonrenewable sources does not meet this requirement.

The jurisdiction shall be provided with an energy analysis as described in Section 506 that documents the renewable energy contribution to the building or a calculation demonstrating that the on-site supply of renewable energy:

1. Is capable of providing at least 3 percent of the total energy load of the building; or

2. Provides on-site renewable energy generation with a nominal (peak) rating of 175 BTU's or 0.50 watts per square foot of building.

506.2.6 Automatic daylighting control system. A minimum of 30 percent of the total conditioned floor area shall be daylight zones with automatic controls that control lights in the daylit areas separately from the non-daylit areas.

Controls for calibration adjustments to the lighting control device shall be accessible to owner authorized personnel. Each daylight control zone shall not exceed 2,500 square feet. Automatic daylighting controls must incorporate an automatic shut-off ability based on time or occupancy in addition to lighting power reduction controls.

Controls will automatically reduce lighting power in response to available daylight by either one of the following methods:

1. Continuous dimming using dimming ballasts and daylight-sensing automatic controls that are capable of reducing the power of general lighting in the daylit zone continuously to less than 35 percent of rated power at maximum light output; or

2. Stepped dimming using multi-level switching and daylight-sensing controls that are capable of reducing lighting power automatically. The system shall provide at least two control channels per zone and be installed in a manner such that at least one control step shall reduce power of general lighting in the daylit zone by 30 to 50 percent of rated power and another control step that reduces lighting power by 65 to 100 percent. Stepped dimming control is not appropriate in continuously occupied areas with ceiling heights of 14 feet or lower.

SECTION 507
TOTAL BUILDING PERFORMANCE

507.1 Scope. This section establishes criteria for compliance using total building performance. The following systems and loads shall be included in determining the total building performance: heating systems, cooling systems, service water heating, fan systems, lighting power, receptacle loads and process loads.

507.2 Mandatory requirements. Compliance with this section requires that the criteria of Sections 502.4, 503.2, 504 and 505 be met.

507.3 Performance-based compliance. Compliance based on total building performance requires that a proposed building (*proposed design*) be shown to have an annual energy cost that is less than or equal to the annual energy cost of the *standard reference design*. Energy prices shall be taken from a source *approved* by the *code official,* such as the Department of Energy, Energy Information Administration's *State Energy Price and Expenditure Report. Code officials* shall be permitted to require time-of-use pricing in energy cost calculations. Nondepletable energy collected off site shall be treated and priced the same as purchased energy. Energy from nondepletable energy sources collected on site shall be omitted from the annual energy cost of the *proposed design.*

> **Exception:** Jurisdictions that require site energy (1 kWh = 3413 Btu) rather than energy cost as the metric of comparison.

507.4 Documentation. Documentation verifying that the methods and accuracy of compliance software tools conform to the provisions of this section shall be provided to the *code official.*

507.4.1 Compliance report. Compliance software tools shall generate a report that documents that the *proposed design* has annual energy costs less than or equal to the annual energy costs of the *standard reference design.* The compliance documentation shall include the following information:

1. Address of the building;
2. An inspection checklist documenting the building component characteristics of the *proposed design* as *listed* in Table 507.5.1(1). The inspection checklist shall show the estimated annual energy cost for both the *standard reference design* and the *proposed design;*
3. Name of individual completing the compliance report; and
4. Name and version of the compliance software tool.

507.4.2 Additional documentation. The *code official* shall be permitted to require the following documents:

1. Documentation of the building component characteristics of the *standard reference design;*
2. Thermal zoning diagrams consisting of floor plans showing the thermal zoning scheme for *standard reference design* and *proposed design;*
3. Input and output report(s) from the energy analysis simulation program containing the complete input and output files, as applicable. The output file shall include energy use totals and energy use by energy source and end-use served, total hours that space conditioning loads are not met and any errors or warning messages generated by the simulation tool as applicable;

4. An explanation of any error or warning messages appearing in the simulation tool output; and
5. A certification signed by the builder providing the building component characteristics of the *proposed design* as given in Table 507.5.1(1).

507.5 Calculation procedure. Except as specified by this section, the *standard reference design* and *proposed design* shall be configured and analyzed using identical methods and techniques.

507.5.1 Building specifications. The *standard reference design* and *proposed design* shall be configured and analyzed as specified by Table 507.5.1(1). Table 507.5.1(1) shall include by reference all notes contained in Table 502.2(1).

507.5.2 Thermal blocks. The *standard reference design* and *proposed design* shall be analyzed using identical thermal blocks as required in Section 507.5.2.1, 507.2.2 or 507.5.2.3.

507.5.2.1 HVAC zones designed. Where HVAC zones are defined on HVAC design drawings, each HVAC *zone* shall be modeled as a separate thermal block.

> **Exception:** Different HVAC zones shall be allowed to be combined to create a single thermal block or identical thermal blocks to which multipliers are applied provided:
>
> 1. The space use classification is the same throughout the thermal block.
> 2. All HVAC zones in the thermal block that are adjacent to glazed exterior walls face the same orientation or their orientations are within 45 degrees (0.79 rad) of each other.
> 3. All of the zones are served by the same HVAC system or by the same kind of HVAC system.

507.5.2.2 HVAC zones not designed. Where HVAC zones have not yet been designed, thermal blocks shall be defined based on similar internal load densities, occupancy, lighting, thermal and temperature schedules, and in combination with the following guidelines:

1. Separate thermal blocks shall be assumed for interior and perimeter spaces. Interior spaces shall be those located more than 15 feet (4572 mm) from an exterior wall. Perimeter spaces shall be those located closer than 15 feet (4572 mm) from an *exterior wall.*
2. Separate thermal blocks shall be assumed for spaces adjacent to glazed exterior walls: a separate *zone* shall be provided for each orientation, except orientations that differ by no more than 45 degrees (0.79 rad) shall be permitted to be considered to be the same orientation. Each *zone* shall include floor area that is 15 feet (4572 mm) or less from a glazed perimeter wall, except that floor area within 15 feet (4572 mm) of glazed perimeter walls having more than one orientation shall be divided proportionately between zones.

3. Separate thermal blocks shall be assumed for spaces having floors that are in contact with the ground or exposed to ambient conditions from zones that do not share these features.

4. Separate thermal blocks shall be assumed for spaces having exterior ceiling or roof assemblies from zones that do not share these features.

507.5.2.3 Multifamily residential buildings. Residential spaces shall be modeled using one thermal block per space except that those facing the same orientations are permitted to be combined into one thermal block. Corner units and units with roof or floor loads shall only be combined with units sharing these features.

507.6 Calculation software tools. Calculation procedures used to comply with this section shall be software tools capable of calculating the annual energy consumption of all building elements that differ between the *standard reference design* and the *proposed design* and shall include the following capabilities.

1. Computer generation of the *standard reference design* using only the input for the *proposed design*. The calculation procedure shall not allow the user to directly modify the building component characteristics of the *standard reference design.*

2. Building operation for a full calendar year (8760 hours).

3. Climate data for a full calendar year (8760 hours) and shall reflect *approved* coincident hourly data for temperature, solar radiation, humidity and wind speed for the building location.

4. Ten or more thermal zones.

5. Thermal mass effects.

6. Hourly variations in occupancy, illumination, receptacle loads, thermostat settings, mechanical ventilation, HVAC equipment availability, service hot water usage and any process loads.

7. Part-load performance curves for mechanical equipment.

8. Capacity and efficiency correction curves for mechanical heating and cooling equipment.

9. Printed *code official* inspection checklist listing each of the *proposed design* component characteristics from Table 507.5.1(1) determined by the analysis to provide compliance, along with their respective performance ratings (e.g., *R*-value, *U*-factor, SHGC, HSPF, AFUE, SEER, EF, etc.).

507.6.1 Specific approval. Performance analysis tools meeting the applicable sub-sections of Section 507 and tested according to ASHRAE Standard 140 shall be permitted to be *approved*. Tools are permitted to be *approved* based on meeting a specified threshold for a jurisdiction. The *code official* shall be permitted to approve tools for a specified application or limited scope.

507.6.2 Input values. When calculations require input values not specified by Sections 502, 503, 504 and 505, those input values shall be taken from an *approved* source.

TABLE 507.5.1(1)
SPECIFICATIONS FOR THE STANDARD REFERENCE AND PROPOSED DESIGNS

BUILDING COMPONENT CHARACTERISTICS	STANDARD REFERENCE DESIGN	PROPOSED DESIGN
Space use classification	Same as proposed	The space use classification shall be chosen in accordance with Table 505.5.2 for all areas of the building covered by this permit. Where the space use classification for a building is not known, the building shall be categorized as an office building.
Roofs	Type: Insulation entirely above deck Gross area: same as proposed U-factor: from Table 502.1.2 Solar absorptance: 0.75 Emittance: 0.90	As proposed As proposed As proposed As proposed As proposed
Walls, above-grade	Type: Mass wall if proposed wall is mass; otherwise steel-framed wall Gross area: same as proposed U-factor: from Table 502.1.2 Solar absorptance: 0.75 Emittance: 0.90	As proposed As proposed As proposed As proposed As proposed
Walls, below-grade	Type: Mass wall Gross area: same as proposed U-Factor: from Table 502.1.2 with insulation layer on interior side of walls	As proposed As proposed As proposed
Floors, above-grade	Type: joist/framed floor Gross area: same as proposed U-factor: from Table 502.1.2	As proposed As proposed As proposed
Floors, slab-on-grade	Type: Unheated F-factor: from Table 502.1.2	As proposed As proposed
Doors	Type: Swinging Area: Same as proposed U-factor: from Table 502.2(1)	As proposed As proposed As proposed
Glazing	Area: 　(a) The proposed glazing area; where the proposed glazing area is less than 40 percent of above-grade wall area. 　(b) 40 percent of above-grade wall area; where the proposed glazing area is 40 percent or more of the above-grade wall area. U-factor: from Table 502.3 SHGC: from Table 502.3 except that for climates with no requirement (NR) SHGC = 0.40 shall be used External shading and PF: None For example, if the proposed building design has a glazing area equal to 45% of the above-grade wall area, the Standard Reference Design shall have a glazing area equal to 30% of the above-grade wall area. On the other hand, if the proposed design has a glazing fraction of 20%, the Standard Reference Design shall also have a glazing fraction of 20%.	As proposed As proposed As proposed As proposed
Skylights	Area: 　(a) The proposed skylight area; where the proposed skylight area is less than 3 percent of gross area of roof assembly. 　(b) 3 percent of gross area of roof assembly; where the proposed skylight area is 3 percent or more of gross area of roof assembly. U-factor: from Table 502.3 SHGC: from Table 502.3 except that for climates with no requirement (NR) SHGC = 0.40 shall be used.	As proposed As proposed As proposed
Lighting, interior	The interior lighting power shall be determined in accordance with Table 505.5.2. Where the occupancy of the building is not known, the lighting power density shall be 1.0 Watt per square foot (10.73 W/m^2) based on the categorization of buildings with unknown space classification as offices.	As proposed
Lighting, exterior	The lighting power shall be determined in accordance with Table 505.6.2(2). Areas and dimensions of tradable and nontradable surfaces shall be the same as proposed.	As proposed

(continued)

BUILDING COMPONENT CHARACTERISTICS	STANDARD REFERENCE DESIGN	PROPOSED DESIGN
Internal gains	Same as proposed	Receptacle, motor and process loads shall be modeled and estimated based on the space use classification. All end-use load components within and associated with the building shall be modeled to include, but not be limited to, the following: exhaust fans, parking garage ventilation fans, exterior building lighting, swimming pool heaters and pumps, elevators, escalators, refrigeration equipment and cooking equipment.
Schedules	Same as proposed	Operating schedules shall include hourly profiles for daily operation and shall account for variations between weekdays, weekends, holidays and any seasonal operation. Schedules shall model the time-dependent variations in occupancy, illumination, receptacle loads, thermostat settings, mechanical ventilation, HVAC equipment availability, service hot water usage and any process loads. The schedules shall be typical of the proposed building type as determined by the designer and approved by the jurisdiction.
Mechanical ventilation	Same as proposed	As proposed, in accordance with Section 503.2.5.
Heating systems	Fuel type: same as proposed design Equipment type[a]: from Tables 507.5.1(2) and 507.5.1(3) Efficiency: from Tables 503.2.3(4) and 503.2.3(5) Capacity[b]: sized proportionally to the capacities in the proposed design based on sizing runs, and shall be established such that no smaller number of unmet heating load hours and no larger heating capacity safety factors are provided than in the proposed design.	As proposed As proposed As proposed As proposed
Cooling systems	Fuel type: same as proposed design Equipment type[c]: from Tables 507.5.1(2) and 507.5.1(3) Efficiency: from Tables 503.2.3(1), 503.2.3(2) and 503.2.3(3) Capacity[b]: sized proportionally to the capacities in the proposed design based on sizing runs, and shall be established such that no smaller number of unmet cooling load hours and no larger cooling capacity safety factors are provided than in the proposed design. Economizer[d]: same as proposed, in accordance with Section 503.4.1.	As proposed As proposed As proposed As proposed As proposed
Service water heating	Fuel type: same as proposed Efficiency: from Table 504.2 Capacity: same as proposed Where no service water hot water system exists or is specified in the proposed design, no service hot water heating shall be modeled.	As proposed As proposed As proposed

a. Where no heating system exists or has been specified, the heating system shall be modeled as fossil fuel. The system characteristics shall be identical in both the standard reference design and proposed design.

b. The ratio between the capacities used in the annual simulations and the capacities determined by sizing runs shall be the same for both the standard reference design and proposed design.

c. Where no cooling system exists or no cooling system has been specified, the cooling system shall be modeled as an air-cooled single-zone system, one unit per thermal zone. The system characteristics shall be identical in both the standard reference design and proposed design.

d. If an economizer is required in accordance with Table 503.3.1 (1), and if no economizer exists or is specified in the proposed design, then a supply air economizer shall be provided in accordance with Section 503.4.1.

TABLE 507.5.1(2)
HVAC SYSTEMS MAP

CONDENSER COOLING SOURCE[a]	HEATING SYSTEM CLASSIFICATION[b]	STANDARD REFERENCE DESIGN HVC SYSTEM TYPE[c]		
		Single-zone Residential System	Single-zone Nonresidential System	All Other
Water/ground	Electric resistance	System 5	System 5	System 1
	Heat pump	System 6	System 6	System 6
	Fossil fuel	System 7	System 7	System 2
Air/none	Electric resistance	System 8	System 9	System 3
	Heat pump	System 8	System 9	System 3
	Fossil fuel	System 10	System 11	System 4

a. Select "water/ground" if the proposed design system condenser is water or evaporatively cooled; select "air/none" if the condenser is air cooled. Closed-circuit dry coolers shall be considered air cooled. Systems utilizing district cooling shall be treated as if the condenser water type were "water." If no mechanical cooling is specified or the mechanical cooling system in the proposed design does not require heat rejection, the system shall be treated as if the condenser water type were "Air." For proposed designs with ground-source or groundwater-source heat pumps, the standard reference design HVAC system shall be water-source heat pump (System 6).

b. Select the path that corresponds to the proposed design heat source: electric resistance, heat pump (including air source and water source), or fuel fired. Systems utilizing district heating (steam or hot water) and systems with no heating capability shall be treated as if the heating system type were "fossil fuel." For systems with mixed fuel heating sources, the system or systems that use the secondary heating source type (the one with the smallest total installed output capacity for the spaces served by the system) shall be modeled identically in the standard reference design and the primary heating source type shall be used to determine *standard reference design HVAC system type.*

c. Select the standard reference design HVAC system category: The system under "single-zone residential system" shall be selected if the HVAC system in the proposed design is a single-zone system and serves a residential space. The system under "single-zone nonresidential system" shall be selected if the HVAC system in the proposed design is a single-zone system and serves other than residential spaces. The system under "all other" shall be selected for all other cases.

TABLE 507.5.1(3)
SPECIFICATIONS FOR THE STANDARD REFERENCE DESIGN HVAC SYSTEM DESCRIPTIONS

SYSTEM NO.	SYSTEM TYPE	FAN CONTROL	COOLING TYPE	HEATING TYPE
1	Variable air volume with parallel fan-powered boxes[a]	VAV[d]	Chilled water[e]	Electric resistance
2	Variable air volume with reheat[b]	VAV[d]	Chilled water[e]	Hot water fossil fuel boiler[f]
3	Packaged variable air volume with parallel fan-powered boxes[a]	VAV[d]	Direct expansion[c]	Electric resistance
4	Packaged variable air volume with reheat[b]	VAV[d]	Direct expansion[c]	Hot water fossil fuel boiler[f]
5	Two-pipe fan coil	Constant volume[i]	Chilled water[e]	Electric resistance
6	Water-source heat pump	Constant volume[i]	Direct expansion[c]	Electric heat pump and boiler[g]
7	Four-pipe fan coil	Constant volume[i]	Chilled water[e]	Hot water fossil fuel boiler[f]
8	Packaged terminal heat pump	Constant volume[i]	Direct expansion[c]	Electric heat pump[h]
9	Packaged rooftop heat pump	Constant volume[i]	Direct expansion[c]	Electric heat pump[h]
10	Packaged terminal air conditioner	Constant volume[i]	Direct expansion	Hot water fossil fuel boiler[f]
11	Packaged rooftop air conditioner	Constant volume[i]	Direct expansion	Fossil fuel furnace

For SI: 1 foot = 304.8 mm, 1 cfm/ft^2 = 0.0004719, 1 Btu/h = 0.293/W, °C = [(°F) -32/1.8].

a. **VAV with parallel boxes:** Fans in parallel VAV fan-powered boxes shall be sized for 50 percent of the peak design flow rate and shall be modeled with 0.35 W/cfm fan power. Minimum volume setpoints for fan-powered boxes shall be equal to the minimum rate for the space required for ventilation consistent with Section 503.4.5, Exception 5. Supply air temperature setpoint shall be constant at the design condition.

b. **VAV with reheat:** Minimum volume setpoints for VAV reheat boxes shall be 0.4 cfm/ft^2 of floor area. Supply air temperature shall be reset based on zone demand from the design temperature to a 10°F temperature difference under minimum load conditions. Design airflow rates shall be sized for the reset supply air temperature, i.e., a 10°F temperature difference.

c. **Direct expansion:** The fuel type for the cooling system shall match that of the cooling system in the proposed design.

d. **VAV:** Constant volume can be modeled if the system qualifies for Exception 1, Section 503.4.5. When the proposed design system has a supply, return or relief fan motor 25 horsepower (hp) or larger, the corresponding fan in the VAV system of the standard reference design shall be modeled assuming a variable speed drive. For smaller fans, a forward-curved centrifugal fan with inlet vanes shall be modeled. If the proposed design's system has a direct digital control system at the zone level, static pressure setpoint reset based on zone requirements in accordance with Section 503.4.2 shall be modeled.

e. **Chilled water:** For systems using purchased chilled water, the chillers are not explicitly modeled and chilled water costs shall be based as determined in Sections 506.3 and 506.5.2. Otherwise, the standard reference design's chiller plant shall be modeled with chillers having the number as indicated in Table 506.5.1(4) as a function of standard reference building chiller plant load and type as indicated in Table 507.5.1(5) as a function of individual chiller load. Where chiller fuel source is mixed, the system in the standard reference design shall have chillers with the same fuel types and with capacities having the same proportional capacity as the proposed design's chillers for each fuel type. Chilled water supply temperature shall be modeled at 44°F design supply temperature and 56°F return temperature. Piping losses shall not be modeled in either building model. Chilled water supply water temperature shall be reset in accordance with Section 503.4.3.4. Pump system power for each pumping system shall be the same as the proposed design; if the proposed design has no chilled water pumps, the standard reference design pump power shall be 22 W/gpm (equal to a pump operating against a 75-foot head, 65-percent combined impeller and motor efficiency). The chilled water system shall be modeled as primary-only variable flow with flow maintained at the design rate through each chiller using a bypass. Chilled water pumps shall be modeled as riding the pump curve or with variable-speed drives when required in Section 503.4.3.4. The heat rejection device shall be an axial fan cooling tower with two-speed fans if required in Section 503.4.4. Condenser water design supply temperature shall be 85°F or 10°F approach to design wet-bulb temperature, whichever is lower, with a design temperature rise of 10°F. The tower shall be controlled to maintain a 70°F leaving water temperature where weather permits, floating up to leaving water temperature at design conditions. Pump system power for each pumping system shall be the same as the proposed design; if the proposed design has no condenser water pumps, the standard reference design pump power shall be 19 W/gpm (equal to a pump operating against a 60-foot head, 60-percent combined impeller and motor efficiency). Each chiller shall be modeled with separate condenser water and chilled water pumps interlocked to operate with the associated chiller.

f. **Fossil fuel boiler:** For systems using purchased hot water or steam, the boilers are not explicitly modeled and hot water or steam costs shall be based on actual utility rates. Otherwise, the boiler plant shall use the same fuel as the proposed design and shall be natural draft. The standard reference design boiler plant shall be modeled with a single boiler if the standard reference design plant load is 600,000 Btu/h and less and with two equally sized boilers for plant capacities exceeding 600,000 Btu/h. Boilers shall be staged as required by the load. Hot water supply temperature shall be modeled at 180°F design supply temperature and 130°F return temperature. Piping losses shall not be modeled in either building model. Hot water supply water temperature shall be reset in accordance with Section 503.4.3.4. Pump system power for each pumping system shall be the same as the proposed design; if the proposed design has no hot water pumps, the standard reference design pump power shall be 19 W/gpm (equal to a pump operating against a 60-foot head, 60-percent combined impeller and motor efficiency). The hot water system shall be modeled as primary only with continuous variable flow. Hot water pumps shall be modeled as riding the pump curve or with variable speed drives when required by Section 503.4.3.4.

g. **Electric heat pump and boiler:** Water-source heat pumps shall be connected to a common heat pump water loop controlled to maintain temperatures between 60°F and 90°F. Heat rejection from the loop shall be provided by an axial fan closed-circuit evaporative fluid cooler with two-speed fans if required in Section 503.4.2. Heat addition to the loop shall be provided by a boiler that uses the same fuel as the proposed design and shall be natural draft. If no boilers exist in the proposed design, the standard reference building boilers shall be fossil fuel. The standard reference design boiler plant shall be modeled with a single boiler if the standard reference design plant load is 600,000 Btu/h or less and with two equally sized boilers for plant capacities exceeding 600,000 Btu/h. Boilers shall be staged as required by the load. Piping losses shall not be modeled in either building model. Pump system power shall be the same as the proposed design; if the proposed design has no pumps, the standard reference design pump power shall be 22 W/gpm, which is equal to a pump operating against a 75-foot head, with a 65-percent combined impeller and motor efficiency. Loop flow shall be variable with flow shutoff at each heat pump when its compressor cycles off as required by Section 503.4.3.3. Loop pumps shall be modeled as riding the pump curve or with variable speed drives when required by Section 503.4.3.4.

h. **Electric heat pump:** Electric air-source heat pumps shall be modeled with electric auxiliary heat. The system shall be controlled with a multistage space thermostat and an outdoor air thermostat wired to energize auxiliary heat only on the last thermostat stage and when outdoor air temperature is less than 40°F.

i. **Constant volume:** Fans shall be controlled in the same manner as in the proposed design; i.e., fan operation whenever the space is occupied or fan operation cycled on calls for heating and cooling. If the fan is modeled as cycling and the fan energy is included in the energy efficiency rating of the equipment, fan energy shall not be modeled explicitly.

TABLE 507.5.1(4)
NUMBER OF CHILLERS

TOTAL CHILLER PLANT CAPACITY	NUMBER OF CHILLERS
≤ 300 tons	1
> 300 tons, < 600 tons	2, sized equally
≥ 600 tons	2 minimum, with chillers added so that no chiller is larger than 800 tons, all sized equally

For SI: 1 ton = 3517 w.

TABLE 507.5.1(5)
WATER CHILLER TYPES

INDIVIDUAL CHILLER PLANT CAPACITY	ELECTRIC CHILLER TYPE	FOSSIL FUEL CHILLER TYPE
≤ 100 tons	Reciprocating	Single-effect absorption, direct fired
> 100 tons, < 300 tons	Screw	Double-effect absorption, direct fired
≥ 300 tons	Centrifugal	Double-effect absorption, direct fired

For SI: 1 ton = 3517 w.

CHAPTER 6
REFERENCED STANDARDS

This chapter lists the standards that are referenced in various sections of this document. The standards are listed herein by the promulgating agency of the standard, the standard identification, the effective date and title, and the section or sections of this document that reference the standard. The application of the referenced standards shall be as specified in Section 107.

AAMA

American Architectural Manufacturers Association
1827 Walden Office Square
Suite 550
Schaumburg, IL 60173-4268

Standard reference number	Title	Referenced in code section number
AAMA/WDMA/CSA 101/I.S.2/A c440—05	Specifications for Windows, Doors and Unit Skylights.	402.4.4, 502.4.1

AHRI

Air Conditioning, Heating, and Refrigeration Institute
4100 North Fairfax Drive
Suite 200
Arlington, VA 22203

Standard reference number	Title	Referenced in code section number
210/240—03	Unitary Air-Conditioning and Air-Source Heat Pump Equipment.	Table 503.2.3(1), Table 503.2.3(2)
310/380—93	Standard for Packaged Terminal Air-conditioners and Heat Pumps.	Table 503.2.3(3)
340/360—2000	Commercial and Industrial Unitary Air-conditioning and Heat Pump Equipment	Table 503.2.3(1), Table 503.2.3(2)
365—02	Commercial and Industrial Unitary Air-conditioning Condensing Units.	Table 503.2.3(6)
440—05	Room Fan-coil.	503.2.8
550/590—98	Water Chilling Packages Using the Vapor Compression Cycle—with Addenda.	Table 503.2.3(7)
560—00	Absorption Water Chilling and Water Heating Packages.	Table 503.2.3(7)
840—1998	Unit Ventilators.	503.2.8
13256-1 (2004)	Water-source Heat Pumps—Testing and Rating for Performance—Part 1: Water-to-air and Brine-to-air Heat Pumps.	Table 503.2.3(2)
1160—2004	Performance Rating of Heat Pump Pool Heaters.	Table 504.2

AMCA

Air Movement and Control Association International
30 West University Drive
Arlington Heights, IL 60004-1806

Standard reference number	Title	Referenced in code section number
500D—07	Laboratory Methods for Testing Dampers for Rating.	502.4.5

ANSI

American National Standards Institute
25 West 43rd Street
Fourth Floor
New York, NY 10036

Standard reference number	Title	Referenced in code section number
Z21.10.3—01	Gas Water Heaters, Volume III - Storage Water Heaters with Input Ratings Above 75,000 Btu per Hour, Circulating Tank and Instantaneous—with Addenda Z21.10.3a-2003 and Z21.10.3b-2004.	Table 504.2
Z21.13—04	Gas-fired Low Pressure Steam and Hot Water Boilers.	Table 503.2.3(5)
Z21.47—03	Gas-fired Central Furnaces.	Table 503.2.3(4)
Z83.8—02	Gas Unit Heaters and Gas-Fired Duct Furnaces—with Addendum Z83.8a-2003.	Table 503.2.3(4)

ASHRAE

American Society of Heating, Refrigerating and Air-Conditioning Engineers, Inc.
1791 Tullie Circle, NE
Atlanta, GA 30329-2305

Standard reference number	Title	Referenced in code section number
119—88 (RA 2004)	Air Leakage Performance for Detached Single-family Residential Buildings	Table 405.5.2(1)
140—2007	Standard Method of Test for the Evaluation of Building Energy Analysis Computer Programs	506.6.1
146—1998	Testing and Rating Pool Heaters	Table 504.2
ANSI/ASHRAE/ACCA Standard 183—2007	Peak Cooling and Heating Load Calculations in Buildings Except Low-rise Residential Buildings	503.2.1
13256-1 (2005)	Water-source Heat Pumps—Testing and Rating for Performance—Part 1: Water-to-air and Brine-to-air Heat Pumps (ANSI/ASHRAE/IESNA 90.1-2004)	Table 503.2.3(2)
90.1—2007	Energy Standard for Buildings Except Low-rise Residential Buildings (ANSI/ASHRAE/IESNA 90.1-2007)	501.1, 501.2, 502.1.1, Table 502.2(2)
ASHRAE—2005	ASHRAE Handbook of Fundamentals	402.1.4, Table 405.5.2(1)
ASHRAE—2004	ASHRAE HVAC Systems and Equipment Handbook-2004	503.2.1

ASME

American Society of Mechanical Engineers
Three Park Avenue
New York, NY 10016-5990

Standard reference number	Title	Referenced in code section number
PTC 4.1 - 1964 (Reaffirmed 1991)	Steam Generating Units	Table 503.2.3(5)

ASTM

ASTM International
100 Barr Harbor Drive
West Conshohocken, PA 19428-2859

Standard reference number	Title	Referenced in code section number
C 90—06b	Specification for Load-bearing Concrete Masonry Units	Table 502.2(1)
E 283—04	Test Method for Determining the Rate of Air Leakage Through Exterior Windows, Curtain Walls and Doors Under Specified Pressure Differences Across the Specimen	402.4.5, 502.4.2, 502.4.8
E 1554—07	Standard Test Methods for Determining Air Leakage of Air Distribution Systems by Fam Pressurization	403.2.2

CSA

Canadian Standards Association
5060 Spectrum Way
Mississauga, Ontario, Canada L4W 5N6

Standard reference number	Title	Referenced in code section number
101/I.S.2/A440—08	Specifications for Windows, Doors and Unit Skylights	402.4.4, 502.4.1

DOE

U.S. Department of Energy
c/o Superintendent of Documents
U.S. Government Printing Office
Washington, DC 20402-9325

Standard reference number	Title	Referenced in code section number
10 CFR Part 430, Subpart B, Appendix E (1998)	Uniform Test Method for Measuring the Energy Consumption of Water Heaters	Table 504.2
10 CFR Part 430, Subpart B, Appendix N (1998)	Uniform Test Method for Measuring the Energy Consumption of Furnaces and Boilers	Table 503.2.3(4), Table 503.2.3(5)

DOE—continued

10 CFR Part 431, Subpart E 2004	Test Procedures and Efficiency Standards for Commercial Packaged Boilers .... Table 503.2.3(6)
DOE/EIA—0376 (Current Edition)	State Energy Prices and Expenditure Report 405.3, 506.2
NAECA—National Appliance Energy Conservation Act	Minimum energy efficiency standards for residential central air conditioners and heat pumps 403.6.2

ICC

International Code Council, Inc.
500 New Jersey Avenue, NW
6th Floor
Washington, DC 20001

Standard reference number	Title	Referenced in code section number
IBC—09	International Building Code®	.201.3, 303.2
IFC—09	International Fire Code®	.201.3
IFGC—09	International Fuel Gas Code®	.201.3
IMC—09	International Mechanical Code®	503.2.5, 503.2.6, 503.2.7.1, 503.2.7.1.1, 503.2.7.1.2, 503.2.9.1, 503.3.1, 503.4.5
IPC—09	International Plumbing Code®	.201.3
IRC—09	International Residential Code®	201.3, 403.2.2, 403.6, 405.6.1, Table 405.5.2(1)

IESNA

Illuminating Engineering Society of North America
120 Wall Street, 17th Floor
New York, NY 10005-4001

Standard reference number	Title	Referenced in code section number
90.1—2007	Energy Standard for Buildings Except Low-rise Residential Buildings	501.1, 501.2, 502.1.1, Table 502.2(2)

NFRC

National Fenestration Rating Council, Inc.
6305 Ivy Lane, Suite 140
Greenbelt, MD 20770

Standard reference number	Title	Referenced in code section number
100—04	Procedure for Determining Fenestration Product U-factors—Second Edition	303.1.3
200—04	Procedure for Determining Fenestration Product Solar Heat Gain Coefficients and Visible Transmittance at Normal Incidence—Second Edition	303.1.3
400—04	Procedure for Determining Fenestration Product Air Leakage—Second Edition	402.4.2, 502.4.1

SMACNA

Sheet Metal and Air Conditioning Contractors National Association, Inc.
4021 Lafayette Center Drive
Chantilly, VA 20151-1209

Standard reference number	Title	Referenced in code section number
SMACNA—85	HVAC Air Duct Leakage Test Manual	503.2.7.1.3

UL

Underwriters Laboratories Inc.
333 Pfingsten Road
Northbrook, IL 60062-2096

Standard reference number	Title	Referenced in code section number
727—06	Oil-fired Central Furnaces	Table 503.2.3(4)
731—95	Oil-fired Unit Heaters—with Revisions through February 2006	Table 503.2.3(4)

US—FTC

United States - Federal Trade Commission
600 Pennsylvania Avenue NW
Washington, DC 20580

Standard reference number	Title	Referenced in code section number
CFR Title 16	R-value Rule.	303.1.4

Window and Door Manufacturers Association
1400 East Touhy Avenue, Suite 470
Des Plaines, IL 60018

Standard reference number	Title	Referenced in code section number
AAMA/WDMA/CSA 101/I.S.2/A440—08	Specifications for Windows, Doors and Unit Skylights.	402.4.4, 502.4.1

APPENDIX 1

RESIDENTIAL REQUIREMENTS

Appendix 1.1. Energy Efficiency Certificate (Section 401.3)

TABLE 401.9
ENERGY EFFICIENCY CERTIFICATE

Builder, Permit Holder or Registered Design Professional Print Name:	
Signature:	
Property Address:	
Date:	
Insulation Rating - List the value covering largest area to all that apply	R-Value
Ceiling/roof:	R-
Wall:	R-
Floor:	R-
Closed Crawl Space Wall:	R-
Closed Crawl Space Floor:	R-
Slab:	R-
Basement Wall:	R-
Fenestration:	
U-Factor	
Solar Heat Gain Coefficient (SHGC)	
Building Air Leakage	
❏ Visually inspected according to 402.4.2.1 OR	
❏ Building Air Leakage Test Results (Sec. 402.4.2.2) ACH50 [Target: 5.0] or CFM50/SFSA [Target: 0.30]	
Name of Tester/Company:	
Date: Phone:	
Ducts:	
Insulation	R-
Total Duct Leakage Test Result (Sect. 403.2.2) (CFM25 Total/100SF) [Target: 6]	
Name of Tester or Company:	
Date: Phone:	
Certificate to be displayed permanently	

APPENDIX 1.2
INSULATION AND AIR SEALING DETAILS

APPENDIX 1.2.1

402.2.1 Ceilings with attic spaces. Exception for fully enclosed attic floor systems

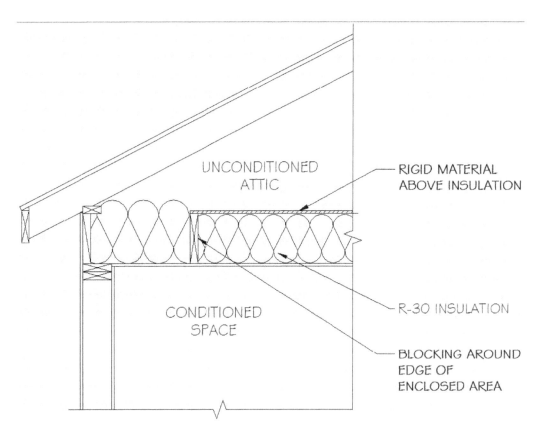

SECTION VIEW OF CEILING WITH ATTIC SPACE

APPENDIX 1.2.2

402.2.9 Closed crawl space walls. Insulation illustrations

Foam or porous insulation has 3" top inspection gap and extends down 3" above top of wall footing or concrete floor

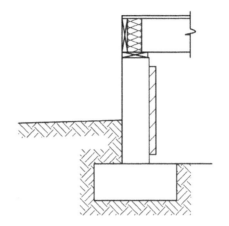

Foam or porous insulation has 3" top inspection gap and extends down 3" above interior ground surface

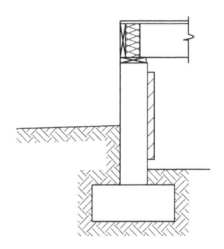

Foam or porous insulation has 3" top inspection gap and extends down 24" below grade

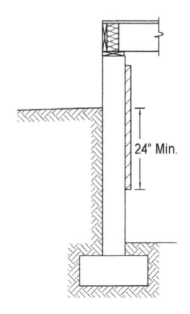

APPENDIX 1.2.3

402.2.12 Framed cavity walls. Insulation enclosure – 1. Tubs

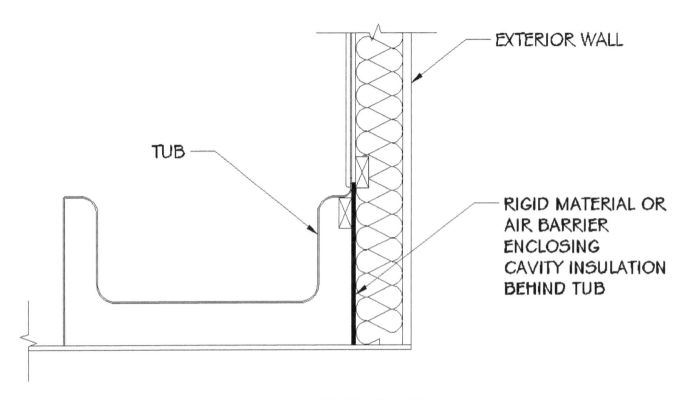

EXTERIOR WALL

TUB

RIGID MATERIAL OR
AIR BARRIER
ENCLOSING
CAVITY INSULATION
BEHIND TUB

SECTION VIEW OF BATH TUB ON EXTERIOR WALL

402.2.12 Framed cavity walls. Insulation enclosure – 2. Showers

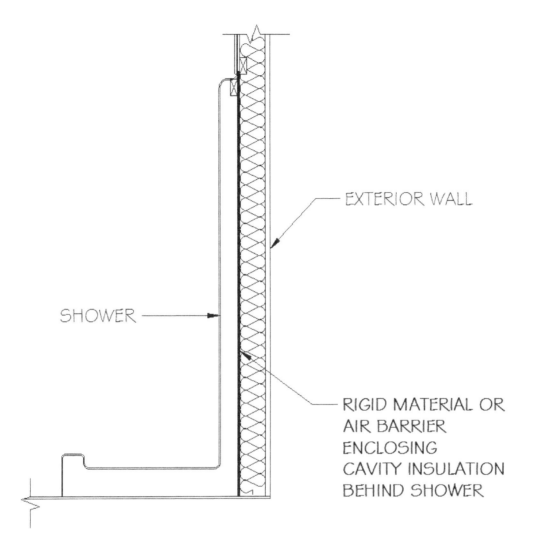

SECTION VIEW OF SHOWER ON EXTERIOR WALL

402.2.12 Framed cavity walls. Insulation enclosure – 3. Stairs

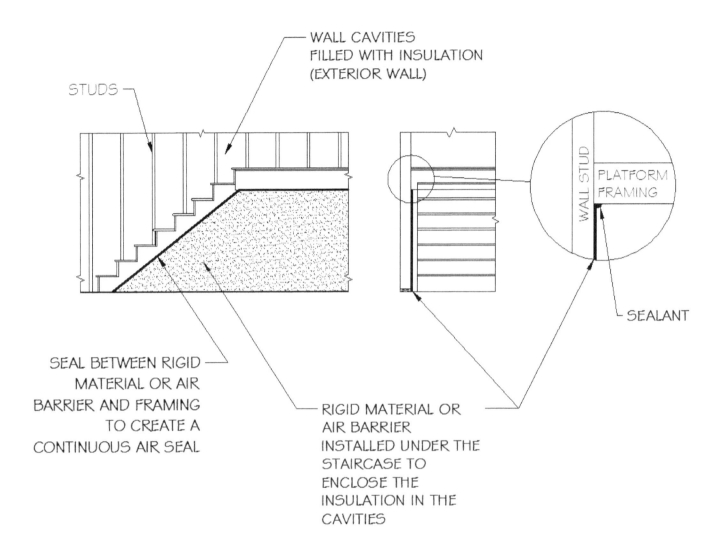

SEAL BETWEEN RIGID MATERIAL OR AIR BARRIER AND FRAMING TO CREATE A CONTINUOUS AIR SEAL

STUDS

WALL CAVITIES FILLED WITH INSULATION (EXTERIOR WALL)

RIGID MATERIAL OR AIR BARRIER INSTALLED UNDER THE STAIRCASE TO ENCLOSE THE INSULATION IN THE CAVITIES

WALL STUD

PLATFORM FRAMING

SEALANT

SECTION VIEW OF INTERIOR STAIRCASE ON EXTERIOR WALL
(OPTION 1)

402.2.12 Framed cavity walls. Insulation enclosure – 3. Stairs

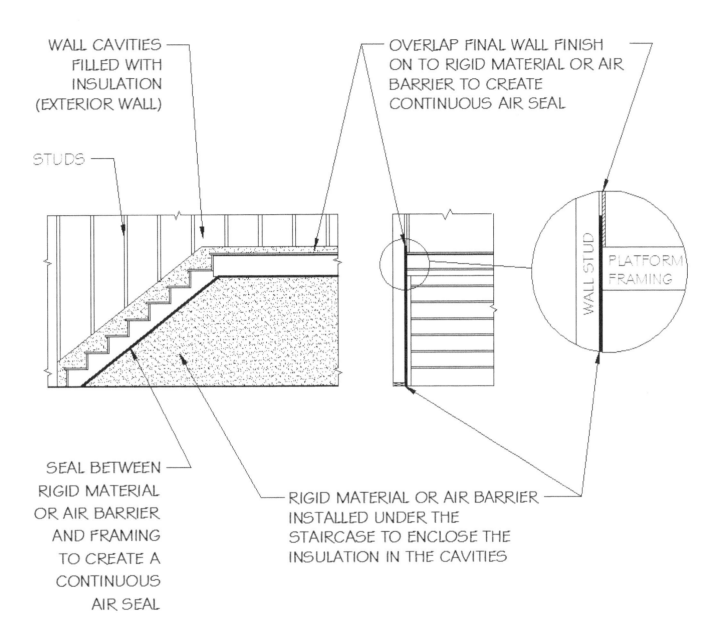

WALL CAVITIES FILLED WITH INSULATION (EXTERIOR WALL)

STUDS

OVERLAP FINAL WALL FINISH ON TO RIGID MATERIAL OR AIR BARRIER TO CREATE CONTINUOUS AIR SEAL

WALL STUD

PLATFORM FRAMING

SEAL BETWEEN RIGID MATERIAL OR AIR BARRIER AND FRAMING TO CREATE A CONTINUOUS AIR SEAL

RIGID MATERIAL OR AIR BARRIER INSTALLED UNDER THE STAIRCASE TO ENCLOSE THE INSULATION IN THE CAVITIES

SECTION VIEW OF INTERIOR STAIRCASE ON EXTERIOR WALL (OPTION 2)

402.2.12 Framed cavity wall. Insulation enclosure – 4. Direct vent gas fireplace

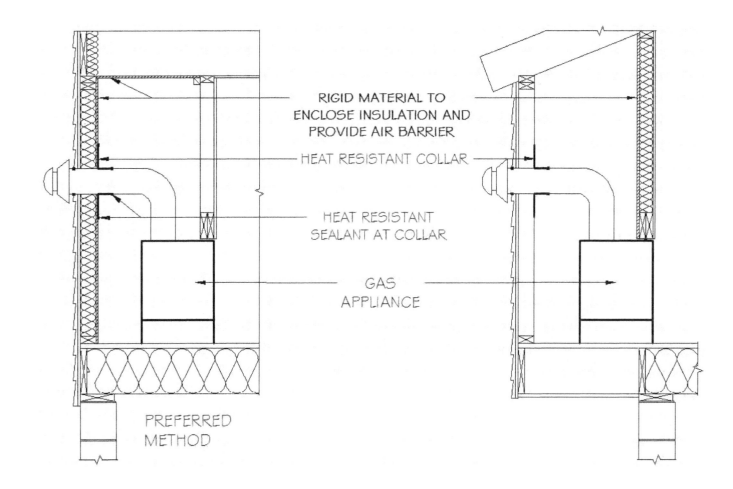

RIGID MATERIAL TO
ENCLOSE INSULATION AND
PROVIDE AIR BARRIER

HEAT RESISTANT COLLAR

HEAT RESISTANT
SEALANT AT COLLAR

GAS
APPLIANCE

PREFERRED
METHOD

SECTION VIEW OF DIRECT VENT GAS FIREPLACE

402.2.12 Framed cavity walls. Insulation enclosure – 5. Walls that adjoin attic spaces

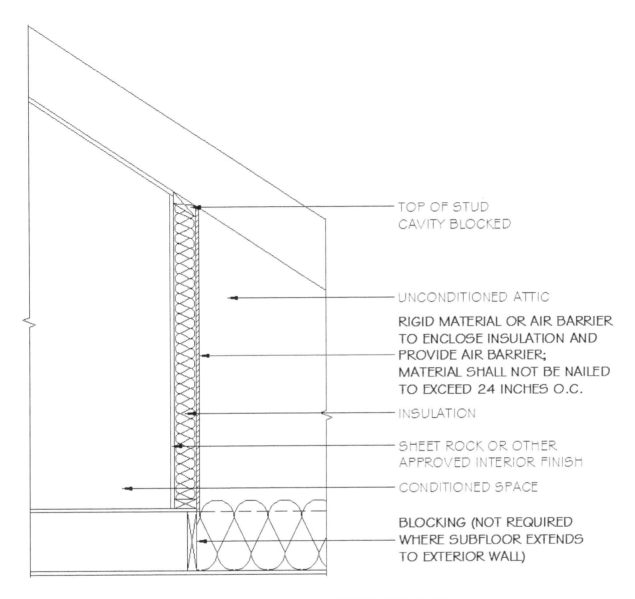

TOP OF STUD
CAVITY BLOCKED

UNCONDITIONED ATTIC

RIGID MATERIAL OR AIR BARRIER
TO ENCLOSE INSULATION AND
PROVIDE AIR BARRIER;
MATERIAL SHALL NOT BE NAILED
TO EXCEED 24 INCHES O.C.

INSULATION

SHEET ROCK OR OTHER
APPROVED INTERIOR FINISH

CONDITIONED SPACE

BLOCKING (NOT REQUIRED
WHERE SUBFLOOR EXTENDS
TO EXTERIOR WALL)

SECTION VIEW OF WALL ADJOINING ATTIC SPACE

402.2.12 Framed cavity walls. Insulation enclosure – 5. Walls that adjoin attic spaces

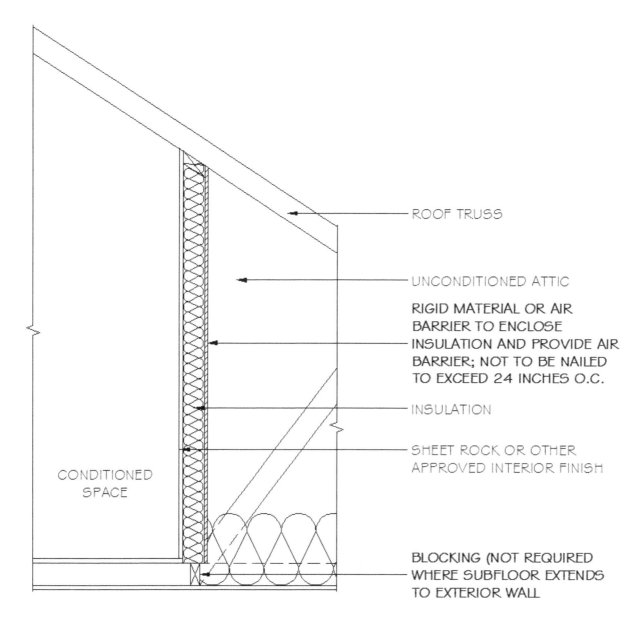

ROOF TRUSS

UNCONDITIONED ATTIC

RIGID MATERIAL OR AIR
BARRIER TO ENCLOSE
INSULATION AND PROVIDE AIR
BARRIER; NOT TO BE NAILED
TO EXCEED 24 INCHES O.C.

INSULATION

SHEET ROCK OR OTHER
APPROVED INTERIOR FINISH

BLOCKING (NOT REQUIRED
WHERE SUBFLOOR EXTENDS
TO EXTERIOR WALL

CONDITIONED
SPACE

SECTION VIEW OF WALL ADJOINING ATTIC SPACE

APPENDIX 1.2.4

402.4.1 Building thermal envelope. – 1. Block and seal floor/ceiling systems

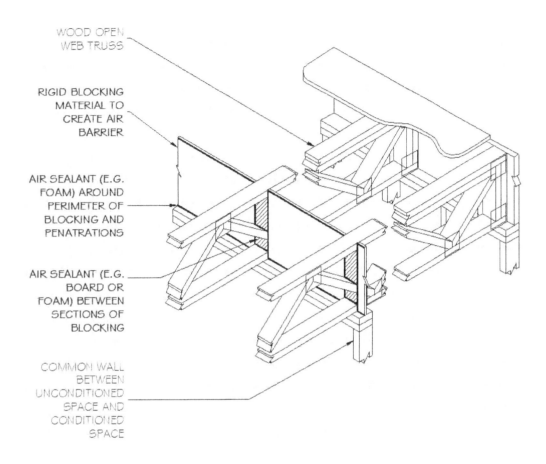

WOOD OPEN
WEB TRUSS

RIGID BLOCKING
MATERIAL TO
CREATE AIR
BARRIER

AIR SEALANT (E.G.
FOAM) AROUND
PERIMETER OF
BLOCKING AND
PENATRATIONS

AIR SEALANT (E.G.
BOARD OR
FOAM) BETWEEN
SECTIONS OF
BLOCKING

COMMON WALL
BETWEEN
UNCONDITIONED
SPACE AND
CONDITIONED
SPACE

**ISOMETRIC VIEW OF DIMENSIONAL LUMBER FLOOR/CEILING SYSTEM
ABOVE COMMON WALL BETWEEN UNCONDITIONED AND CONDITIONED SPACE**

402.4.1 Building thermal envelope. – 1. Block and seal floor/ceiling systems

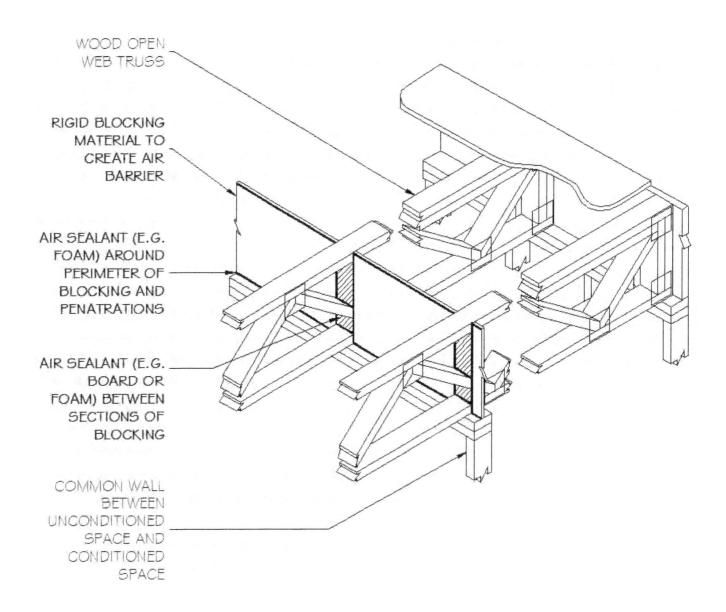

WOOD OPEN
WEB TRUSS

RIGID BLOCKING
MATERIAL TO
CREATE AIR
BARRIER

AIR SEALANT (E.G.
FOAM) AROUND
PERIMETER OF
BLOCKING AND
PENATRATIONS

AIR SEALANT (E.G.
BOARD OR
FOAM) BETWEEN
SECTIONS OF
BLOCKING

COMMON WALL
BETWEEN
UNCONDITIONED
SPACE AND
CONDITIONED
SPACE

**ISOMETRIC VIEW OF WOOD TRUSS FLOOR/CEILING SYSTEM
ABOVE COMMON WALL BETWEEN UNCONDITIONED AND CONDITIONED SPACE**

402.4.1 Building thermal envelope. – 1. Block and seal floor/ceiling systems

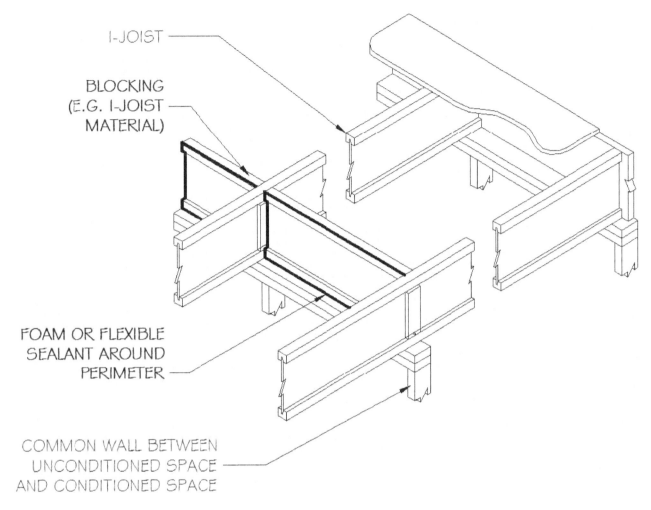

I-JOIST

BLOCKING
(E.G. I-JOIST
MATERIAL)

FOAM OR FLEXIBLE
SEALANT AROUND
PERIMETER

COMMON WALL BETWEEN
UNCONDITIONED SPACE
AND CONDITIONED SPACE

**ISOMETRIC VIEW OF I-JOIST FLOOR/CEILING SYSTEM
ABOVE COMMON WALL BETWEEN UNCONDITIONED AND CONDITIONED SPACE**

402.4.1 Building thermal envelope. – 2. Cap and seal shafts and chases

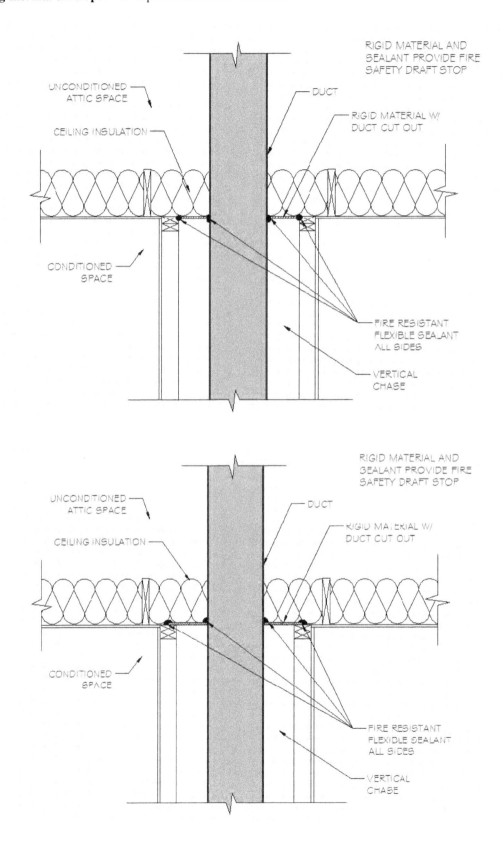

SECTION VIEWS OF DUCT PENETRATING INTO ATTIC

402.4.1 Building thermal envelope. – 3. Cap and seal soffit or dropped ceiling

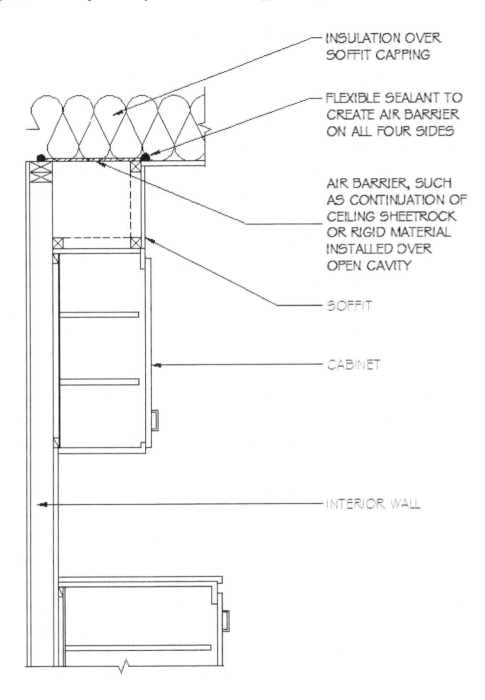

INSULATION OVER
SOFFIT CAPPING

FLEXIBLE SEALANT TO
CREATE AIR BARRIER
ON ALL FOUR SIDES

AIR BARRIER, SUCH
AS CONTINUATION OF
CEILING SHEETROCK
OR RIGID MATERIAL
INSTALLED OVER
OPEN CAVITY

SOFFIT

CABINET

INTERIOR WALL

SECTION VIEW OF SOFFIT OVER CABINET

402.4.1 Building thermal envelope. – 4. Seal HVAC boot penetration – floor

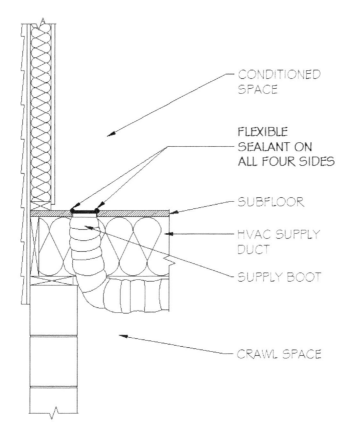

SECTION VIEW OF FLOOR HVAC BOOT PENETRATION

402.4.1 Building thermal envelope. – 4. Seal HVAC boot penetration – ceiling

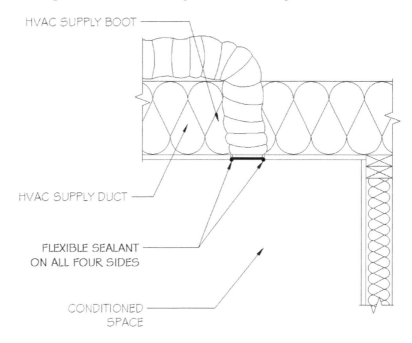

SECTION VIEW OF CEILING HVAC BOOT PENETRATION

APPENDIX 2
COMMERCIAL BUILDING REQUIREMENTS

APPENDIX 2.1: Air Sealing Details

502.4.3 Sealing of the building envelope. - 1. Air seal around fenestration frames

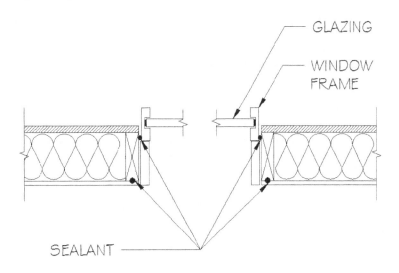

PLAN VIEW OF WINDOW IN WOOD FRAME WALL

502.4.3 Sealing of the building envelope. - 1. Air seal around fenestration frames

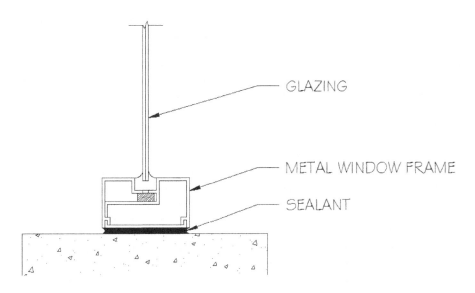

SECTION VIEW OF CURTAIN WALL

502.4.3 Sealing of the building envelope. - 1. Air seal around door frames

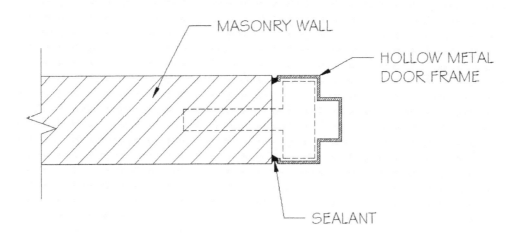

PLAN VIEW OF HOLLOW METAL DOOR FRAME

502.4.3 Sealing of the building envelope. - 2. Air seal junctions between walls and foundations

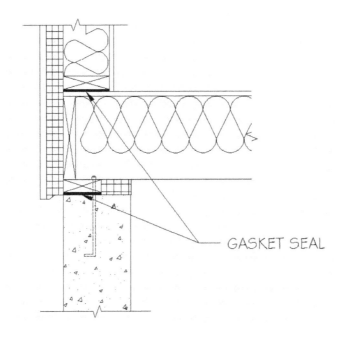

SECTION VIEW OF WOOD FRAME WALL, FLOOR, AND FOUNDATION

502.4.3 Sealing of the building envelope. - 2. Air seal junctions between walls and foundations

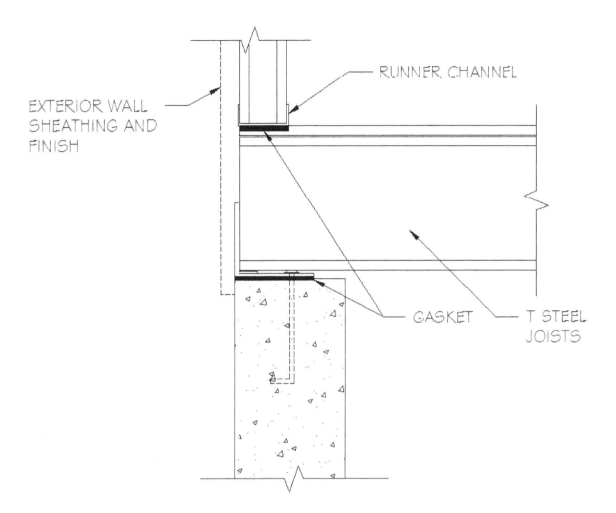

EXTERIOR WALL
SHEATHING AND
FINISH

RUNNER CHANNEL

GASKET

T STEEL
JOISTS

SECTION VIEW OF METAL STUD WALL, FLOOR, AND FOUNDATION

502.4.3 Sealing of the building envelope. - 2. Air seal between walls and roof

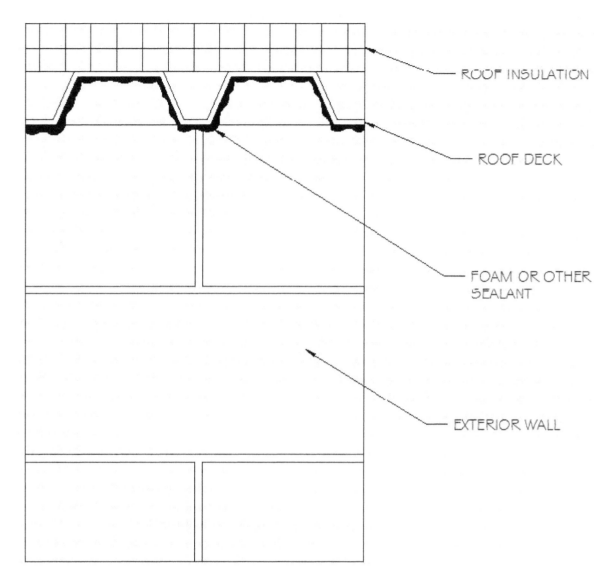

ROOF INSULATION

ROOF DECK

FOAM OR OTHER
SEALANT

EXTERIOR WALL

SECTION VIEW (A) OF WALL AND ROOF

502.4.3 Sealing of the building envelope. - 2. Air seal between walls and roof

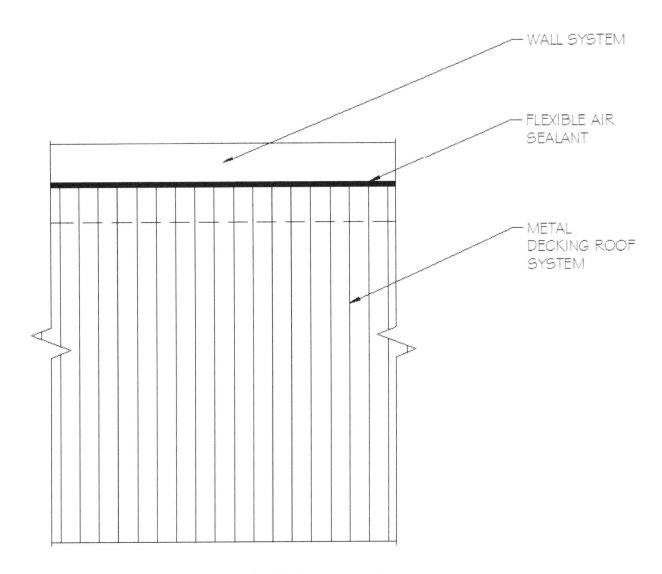

WALL SYSTEM

FLEXIBLE AIR SEALANT

METAL DECKING ROOF SYSTEM

PLAN VIEW (B) OF WALL AND ROOF

502.4.3 Sealing of the building envelope. - 3. Air seal utility services through walls

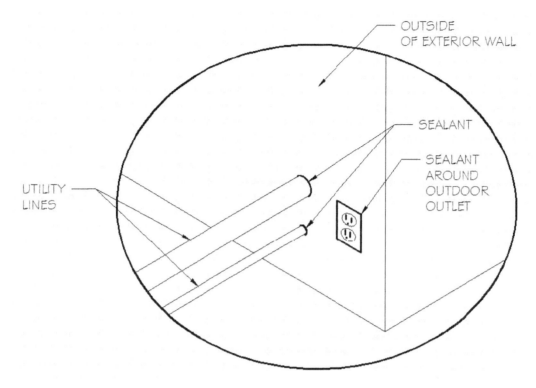

SEAL EXTERIOR PENETRATIONS

502.4.3 Sealing of the building envelope. - 3. Air sealing utility services through walls.

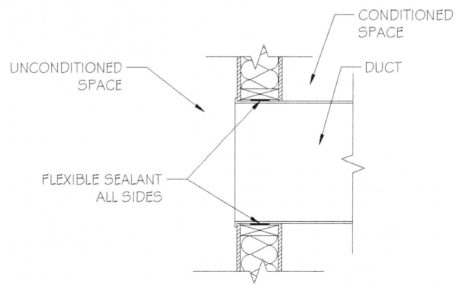

PLAN VIEW OF WALL HVAC DUCT PENETRATING EXTERIOR WALL

502.4.3 Sealing of the building envelope. - 3. Air sealing utility services through walls and floors.

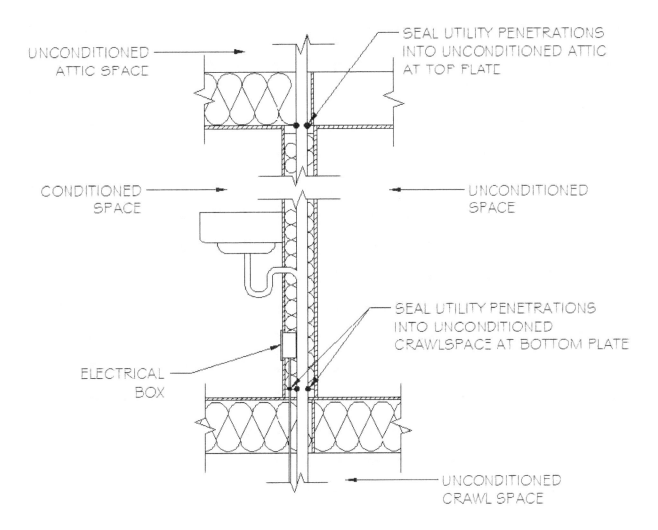

SECTION VIEW OF WALL AND SEALING POINTS OF UTILITY PENETRATIONS

502.4.3 Sealing of the building envelope.

5. Air sealing joints, seams, and penetrations of the air barrier system. Compliant systems include:

A. OSB or plywood sheathing covered with continuous 15-pound felt or other water and vapor resistant treatment with joints lapped 6 inches minimum

B. Building wrap with sealed seams

C. Drywall with sealed joints that completely covers the interior or exterior side of all exterior framed wall cavities with the exception of interior wall intersections (see next page)

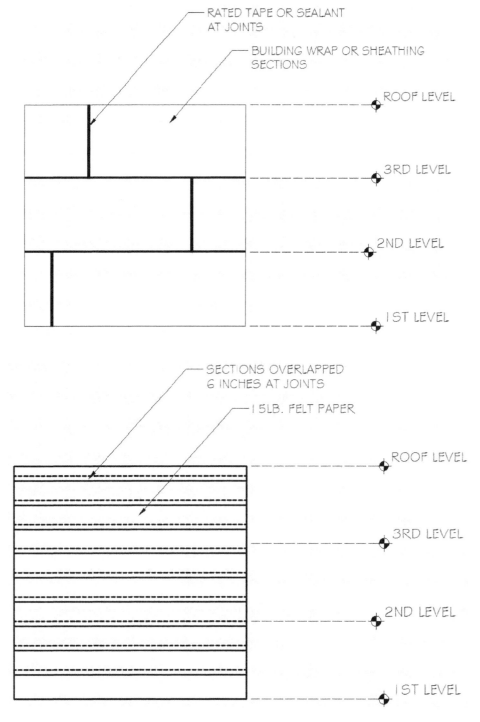

ELEVATION VIEW OF A 3 STORY COMMERCIAL BUILDING

502.4.3 Sealing of the building envelope. - 6. Other openings in the building envelope, continuous wall enclosure

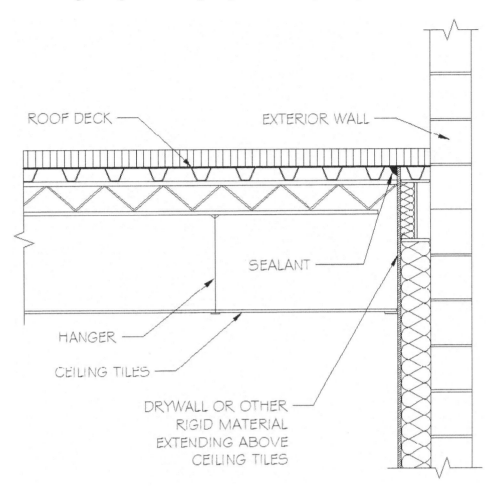

SECTION VIEW OF SUSPENDED CEILING TILES AND EXTERIOR WALL

APPENDIX 2.2: Commercial Building Opaque Assemblies
Table 502.2(2) Roofs.

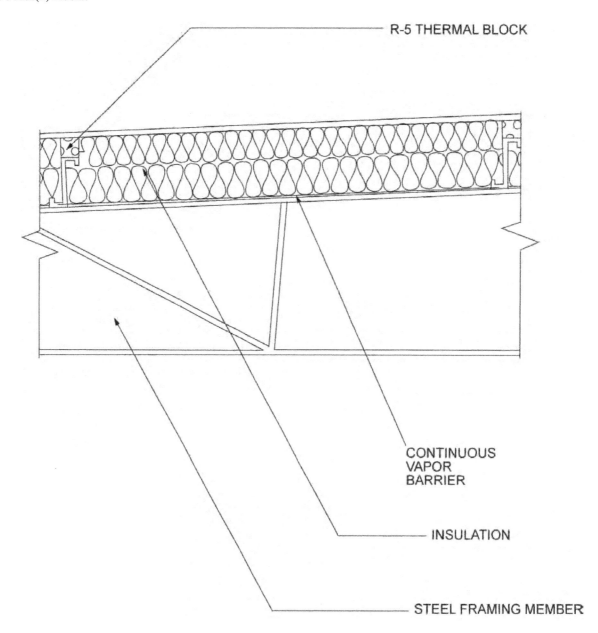

SECTION VIEW OF METAL FRAME ROOF:
FILLED CAVITY FIBERGLASS INSULATION

Table 502.2(2) Roofs.

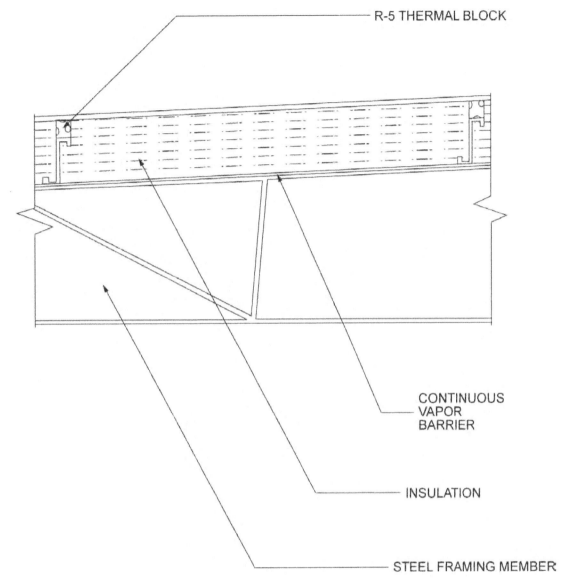

SECTION OF METAL FRAME ROOF:
LINER SYSTEM WITH MINIMUM <u>R-5 THERMAL BLOCK</u>

Table 502.2(2) Walls.

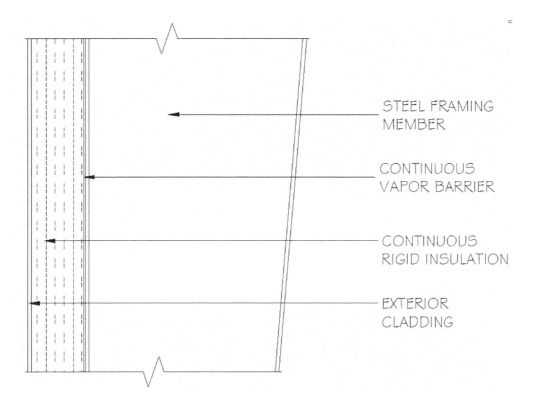

STEEL FRAMING
MEMBER

CONTINUOUS
VAPOR BARRIER

CONTINUOUS
RIGID INSULATION

EXTERIOR
CLADDING

SECTION VIEW OF METAL FRAME WITH INSULATION BETWEEN FRAME AND EXTERIOR PANEL

APPENDIX 3

SAMPLE WORKSHEETS FOR
RESIDENTIAL AIR AND DUCT LEAKAGE TESTING

APPENDIX 3A
Air sealing: Visual inspection option (Section 402.4.2.1)
Sample Worksheet

402.4.2 Air sealing. Building envelope air tightness shall be demonstrated by Section 402.4.2.1 or 402.4.2.2.

402.4.2.1 Visual inspection option. Building envelope tightness shall be considered acceptable when items providing insulation enclosure in Section 402.2.12 and air sealing in Section 402.4.1 are addressed and when the items listed in Table 402.4.2, applicable to the method of construction, are certified by the by the builder, permit holder or registered design professional via the certificate in Appendix 1.1.

TABLE 402.4.2
AIR BARRIER INSPECTION

COMPONENT	CRITERIA
Ceiling/attic	Sealants or gaskets provide a continuous air barrier system joining the top plate of framed walls with either the ceiling drywall or the top edge of wall drywall to prevent air leakage. Top plate penetrations are sealed.
	For ceiling finishes that are not air barrier systems such as tongue-and-groove planks, air barrier systems,(for example, taped house wrap), shall be used above the finish
	Note: It is acceptable that sealants or gaskets applied as part of the application of the drywall will not be observable by the code official
Walls	Sill plate is gasketed or sealed to subfloor or slab.
Windows and doors	Space between window and exterior door jambs and framing is sealed.
Floors (including above-garage and cantilevered floors)	Air barrier system is installed at any exposed edge of insulation.
Penetrations	Utility penetrations through the building thermal envelope, including those for plumbing, electrical wiring, ductwork, security and fire alarm wiring, and control wiring, shall be sealed.
Garage separation	Air sealing is provided between the garage and conditioned spaces. An air barrier system shall be installed between the ceiling system above the garage and the ceiling system of interior spaces.
Duct boots	Sealing HVAC register boots and return boxes to subfloor or drywall.
Recessed lighting	Recessed light fixtures are air tight, IC rated, and sealed to drywall.
	Exception—fixtures not penetrating the building envelope.

Property Address:

_____

402.4.2.1 Visual inspection option

The inspection information including tester name, date, and contact shall be included on the certificate described in Section 401.3.

_____   _____

Signature   Date

APPENDIX 3B
Air sealing: Testing option (Section 402.4.2.2)
Sample Worksheet

402.4.2 Air sealing. Building envelope air tightness shall be demonstrated by Section 402.4.2.1 or 402.4.2.2.

402.4.2.2 Testing option. Building envelope tightness shall be considered acceptable when items providing insulation enclosure in Section 402.2.12 and air sealing in Section 402.4.1 are addressed and when tested air leakage is less than or equal to one of the two following performance measurements:

    1. 0.30 CFM50/Square Foot of Surface Area (SFSA) or

    2. Five (5) air changes per hour (ACH50)

When tested with a blower door fan assembly, at a pressure of 33.5 psf (50 Pa). A single point depressurization, not temperature corrected, test is sufficient to comply with this provision, provided that the blower door fan assembly has been certified by the manufacturer to be capable of conducting tests in accordance with ASTM E 779-03. Testing shall occur after rough in and after installation of penetrations of the building envelope, including penetrations for utilities, plumbing, electrical, ventilation and combustion appliances. Testing shall be reported by the permit holder, a NC licensed general contractor, a NC licensed HVAC contractor, a NC licensed Home Inspector, a registered design professional, *a certified BPI Envelope Professional or a certified HERS rater.*

During testing:

    1. Exterior windows and doors, fireplace and stove doors shall be closed, but not sealed;

    2. Dampers shall be closed, but not sealed, including exhaust, backdraft, and flue dampers;

    3. Interior doors shall be open;

    4. Exterior openings for continuous ventilation systems, air intake ducted to the return side of the conditioning system, and energy or heat recovery ventilators shall be closed and sealed;

    5. Heating and cooling system(s) shall be turned off; and

    6. Supply and return registers shall not be sealed.

The air leakage information, including building air leakage result, tester name, date, and contact information, shall be included on the certificate described in Section 401.3.

For Test Criteria 1 above, the report shall be produced in the following manner: Perform the blower door test and record the *CFM50*_____. Calculate the total square feet of surface area for the building thermal envelope, all floors, ceilings, and walls (this includes windows and doors) and record the area_____. Divide *CFM50* by the total square feet and record the result below. If the result is less than or equal to **[0.24 CFM50/SFSA]** the envelope tightness is acceptable; or

For Test Criteria 2 above, the report shall be produced in the following manner: Perform a blower door test and record the *CFM50*_____. Multiply the CFM50 by 60 minutes to create CFHour50 and record _____. Then calculate the total conditioned volume of the home and record_____. Divide the CFH50 by the total volume and record the result below. If the result is less than or equal to **[4 ACH50]** the envelope tightness is acceptable.

Property Address: _____

Fan attachment location: _____Company Name: _____

Contact Information: _____

_____   _____
Signature of Tester   Date

Permit Holder, NC Licensed General Contractor, NC Licensed HVAC Contractor, NC Licensed Home Inspector,
Registered Design Professional, Certified BPI Envelope Professional, or Certified HERS Rater
(circle one)

APPENDIX 3C
Duct sealing: Duct air leakage test (Section 403.2.2)
Sample Worksheet

N1103.2.2 Sealing. All ducts, air handlers, filter boxes and building cavities used as ducts shall be sealed. Joints and seams shall comply with Section 603 of the *NC Mechanical Code.*

Duct tightness shall be verified as follows:

Total duct leakage less than or equal to 6 CFM (18 L/min) per 100 ft² (9.29 m²) of *conditioned floor area* served by that system when tested at a pressure differential of 0.1 inches w.g. (25 Pa) across the entire system, including the manufacturer's air handler enclosure.

During testing:

1. Block, if present, the ventilation air duct connected to the conditioning system.
2. The duct air leakage testing equipment shall be attached to the largest return in the system or to the air handler.
3. The filter shall be removed and the air handler power shall be turned off.
3. Supply boots or registers and return boxes or grilles shall be taped, plugged, or otherwise sealed air tight.
4. The hose for measuring the 25 Pascals of pressure differential shall be inserted into the boot of the supply that is nominally closest to the air handler.
5. Specific instructions from the duct testing equipment manufacturer shall be followed to reach duct test pressure and measure duct air leakage.

Testing shall be performed and reported by the permit holder, a NC licensed general contractor, a NC licensed HVAC contractor, a NC licensed Home Inspector, a registered design professional, a certified BPI Envelope Professional or a certified HERS rater. A single point depressurization, not temperature corrected, test is sufficient to comply with this provision, provided that the duct testing fan assembly has been certified by the manufacturer to be capable of conducting tests in accordance with ASTM E1554-07.

The duct leakage information, including duct leakage result, tester name, date, company and contact information, shall be included on the certificate described in Section 401.3.

For the Test Criteria, the report shall be produced in the following manner: perform the HVAC system air leakage test and record the CFM25. Calculate the total square feet of Conditioned Floor Area (CFA) served by that system. Multiply CFM25 by 100, divide the result by the CFA and record the result. If the result is less than or equal to [6 CFM25/100 SF] the HVAC system air tightness is acceptable.

Complete one duct leakage report for each HVAC system serving the home:

Property Address: _____

HVAC System Number: _____ Describe area of home served: _____

CFM25 Total _____. Conditioned Floor Area (CFA) served by system: _____ s.f.

CFM25 x 100 divided by CFA = _____ CFM25/100SF (e.g. 100 CFM25 x 100/2,000 CFA = 5 CFM25/100SF)

Fan attachment location: _____

Company Name: _____

Contact Information: _____

_____   _____
Signature of Tester   Date

Permit Holder, NC Licensed General Contractor, NC Licensed HVAC Contractor, NC Licensed Home Inspector,
Registered Design Professional, Certified BPI Envelope Professional, or Certified HERS Rater

(circle one)

ADDITIONAL VOLUNTARY CRITERIA FOR INCREASING ENERGY EFFICIENCY (High Efficiency Residential Option)

1. **Introduction.** The increased energy efficiency measures identified in this appendix are strictly voluntary at the option of the permit holder and have been evaluated to be the most cost effective measures for achieving an additional 15-20% energy efficiency beyond the code minimums.

2. **Requirements.** Follow all sections of the Chapter 4 of the *North Carolina Energy Conservation Code*, except the following.

   a. Instead of using Table 402.1.1 in Section 402.1.1, use Table 4A shown below.

TABLE 4A
INSULATION AND FENESTRATION REQUIREMENTS BY COMPONENT[a]

CLIMATE ZONE	FENESTRATION U-FACTOR[b]	SKYLIGHT[b] U-FACTOR	GLAZED FENESTRATION SHGC[b, e]	CEILING R-VALUE[k]	WOOD FRAME WALL R-VALUE[e]	MASS WALL R-VALUE[i]	FLOOR R-VALUE	BASEMENT[c] WALL R-VALUE	SLAB[d] R-VALUE	CRAWL SPACE[c] WALL R-VALUE
3	0.32[j]	0.65	0.25	38	19, 13+5, or 15+3[e,h]	5/10	19	10/13[f]	5	10/13
4	0.32	0.60	0.25	38	19, 13+5, or 15+3[e,h]	5/10	19	10/13	10	10/13
5	0.32	0.60	(NR)	38	19, 13+5, or 15+3[e,h]	13/17	30[g]	10/13	10	15/19

For SI: 1 foot = 304.8 mm.

a. R-values are minimums. U-factors and SHGC are maximums.

b. The fenestration U-factor column excludes skylights. The SHGC column applies to all glazed fenestration.

c. "10/13" means R-10 continuous insulated sheathing on the interior or exterior of the home or R-13 cavity insulation at the interior of the basement wall or crawl space wall.

d. For monolithic slabs, insulation shall be applied from the inspection gap downward to the bottom of the footing or a maximum of 18 inches below grade, whichever is less . For floating slabs, insulation shall extend to the bottom of the foundation wall or 24 inches, whichever is less. (See Appendix O) R-5 shall be added to the required slab edge R-values for heated slabs.

e. R-19 fiberglass batts compressed and installed in a nominal 2 × 6 framing cavity is deemed to comply. Fiberglass batts rated R-19 or higher compressed and installed in a 2 × 4 wall is not deemed to comply.

f. Basement wall insulation is not required in warm-humid locations as defined by Figure 301.2 and Tables 301.1 and 301.3.

g. Or insulation sufficient to fill the framing cavity, R-19 minimum.

h. "13+5" means R-13 cavity insulation plus R-5 insulated sheathing. 15+3 means R-15 cavity insulation plus R-3 insulated sheathing. If structural sheathing covers 25 percent or less of the exterior, insulating sheathing is not required where structural sheathing is used. If structural sheathing covers more than 25 percent of exterior, structural sheathing shall be supplemented with insulated sheathing of at least R-2. 13+2.5 means R-13 cavity insulation plus R-2.5 sheathing.

i. For Mass Walls, the second R-value applies when more than half the insulation is on the interior of the mass wall.

j. R-30 shall be deemed to satisfy the ceiling insulation requirement wherever the full height of uncompressed R-30 insulation extends over the wall top plate at the eaves. Otherwise R-38 insulation is required where adequate clearance exists or insulation must extend to either the insulation baffle or within 1" of the attic roof deck.

k. Table value required except for roof edge where the space is limited by the pitch of the roof, there the insulation must fill the space up to the air baffle.

   b. Instead of using Table 402.1.3 in Section 402.1.3, use Table 4B to find the maximum U-factors for building components.

TABLE 4B
EQUIVALENT U-FACTORS[a]

CLIMATE ZONE	FENESTRATION U-FACTOR	SKYLIGHT U-FACTOR	CEILING U-FACTOR	FRAME WALL U-FACTOR	MASS WALL U-FACTOR[b]	FLOOR U-FACTOR	BASEMENT WALL U-FACTOR[d]	CRAWL SPACE WALL U-FACTOR[c]
3	0.32	0.65	0.030	0.061	0.141	0.047	0.059	0.065
4	0.32	0.60	0.030	0.061	0.141	0.047	0.059	0.065
5	0.32	0.60	0.030	0.061	0.082	0.033	0.059	0.055

a. Nonfenestration U-factors shall be obtained from measurement, calculation or an approved source.

b. When more than half the insulation is on the interior, the mass wall U-factors shall be a maximum 0.12 in Zone 3, 0.10 in Zone 4, and the same as the frame wall U-factor in Zone 5.

c. Basement wall U-factor of 0.360 in warm-humid locations as defined by Table 301.1 and Figure 301.2.

d. Foundation U-factor requirements shown in Table 4B include wall construction and interior air films but exclude soil conductivity and exterior air films. U-factors for determining code compliance in accordance with Section 402.1.4 (total UA alternative) shall be modified to include soil conductivity and exterior air films.

   c. Instead of using the air leakage value for maximum leakage shown in Section 402.4.2.2, use the following:

      i. 0.24 CFM50/Square Foot of Surface Area (SFSA) or

ii. Four (4) air changes per hour (ACH50)

d. Instead of using the duct leakage value for maximum leakage shown in Section 403.2.2, use the following:

Total duct leakage less than or equal to 4 CFM (12 L/min) per 100 ft2 (9.29 m^2) of *conditioned floor area* served by that system when tested at a pressure differential of 0.1 inches w.g. (25 Pa) across

Table 4C
Sample Confirmation Form for
ADDITIONAL VOLUNTARY CRITERIA FOR INCREASING ENERGY EFFICIENCY
(High Efficiency Residential Option)

NORTH CAROLINA ENERGY CONSERVATION CODE: HIGH EFFICIENCY RESIDENTIAL OPTION INSULATION AND FENESTRATION VALUES				PROPOSED PROJECT VALUES
Climate Zone	**3**	**4**	**5**	
Fenestration *U*-Factor	0.32^j	0.32^j	0.32^j	
Skylight *U*-Factor	0.65	0.6	0.6	
Glazed Fenestration SHGC$^{b, e}$	0.25	0.25	NR	
Ceiling *R*-value	38	38	38	
Wood Frame Wall *R*-value^e	19, 13+5, or 15+3^{eh}	19, 13+5, or 15+3^{eh}	19, 13+5, or 15+3^{eh}	
Mass Wall *R*-value^j	5/10	5/10	13/17	
Floor *R*-value	19	19	30^g	
Basement Wall *R*-value^c	10/13^f	10/13^f	10/13^f	
Slab *R*-value and Depth^d	5, 2 ft	10, 2 ft	10, 2 ft	
Crawl Space Wall *R*-value^c	10/13	10/13	15/19	
Building Air Leakage				
Visually inspected according to N1102.4.2.1 (check box) OR				
Building Air Leakage Test according to N1102.4.2.2 (check box). Show test value:				
ACH50 [Target: 4.0], or				
CFM50/SFSA [Target: 0.24]				
Name of Tester / Company:				
Date:				
Duct Insulation and Sealing				
Insulation Value	R-			
Duct Leakage Test Result (Sect. N1103.2.2)				
(CFM25 Total/100SF) [Target:4]				
Name of Tester or Company:				
Date:				

4D:
SAMPLE WORKSHEETS FOR RESIDENTIAL AIR AND DUCT LEAKAGE TESTING
4D.1
Air sealing: Visual inspection option (Section 402.4.2.1)
Sample Worksheet for Alternative Residential Energy Code for Higher Efficiency

402.4.2 Air sealing. Building envelope air tightness shall be demonstrated by Section 402.4.2.1 or 402.4.2.2:

402.4.2.1 Visual inspection option. Building envelope tightness shall be considered acceptable when items providing insulation enclosure in Section 402.2.12 and air sealing in Section 402.4.1 are addressed and when the items listed in Table 402.4.2, applicable to the method of construction, are certified by the builder, permit holder or registered design professional via the certificate in Appendix 1.1.

TABLE 402.4.2
AIR BARRIER INSPECTION

COMPONENT	CRITERIA
Ceiling/attic	Sealants or gaskets provide a continuous air barrier system joining the top plate of framed walls with either the ceiling drywall or the top edge of wall drywall to prevent air leakage. Top plate penetrations are sealed. For ceiling finishes that are not air barrier systems such as tongue-and-groove planks, air barrier systems; (for example, taped house wrap), shall be used above the finish. **Note:** It is acceptable that sealants or gaskets applied as part of the application of the drywall will not be observable by the code official
Walls	Sill plate is gasketed or sealed to subfloor or slab.
Windows and doors	Space between window and exterior door jambs and framing is sealed.
Floors (including above-garage and cantilevered floors)	Air barrier system is installed at any exposed edge of insulation.
Penetrations	Utility penetrations through the building thermal envelope, including those for plumbing, electrical wiring, ductwork, security and fire alarm wiring, and control wiring, shall be sealed.
Garage separation	Air sealing is provided between the garage and conditioned spaces. An air barrier system shall be installed between the ceiling system above the garage and the ceiling system of interior spaces.
Duct boots	Sealing HVAC register boots and return boxes to subfloor or drywall.
Recessed lighting	Recessed light fixtures are air tight, IC rated, and sealed to drywall. **Exception**—fixtures not penetrating the building envelope.

Property Address:

_____

402.4.2.1 Visual inspection option

The inspection information including tester name, date, and contact shall be included on the certificate described in Section 401.3.

_____   _____

Signature   Date

4D.2
Air sealing: Testing option (Section 402.4.2.2)
Sample Worksheet for Alternative Residential Energy Code for Higher Efficiency

402.4.2 Air sealing. Building envelope air tightness shall be demonstrated by Section 402.4.2.1 or 402.4.2.2:

402.4.2.2 Testing option. Building envelope tightness shall be considered acceptable when items providing insulation enclosure in Section 402.2.12 and air sealing in Section 402.4.1 are addressed and when tested air leakage is less than or equal to one of the two following performance measurements:

    1. 0.24 CFM50/Square Foot of Surface Area (SFSA) or

    2. Four (4) air changes per hour (ACH50)

When tested with a blower door fan assembly, at a pressure of 33.5 psf (50 Pa). A single point depressurization, not temperature corrected, test is sufficient to comply with this provision, provided that the blower door fan assembly has been certified by the manufacturer to be capable of conducting tests in accordance with ASTM E 779-03. Testing shall occur after rough in and after installation of penetrations of the building envelope, including penetrations for utilities, plumbing, electrical, ventilation and combustion appliances. Testing shall be reported by the permit holder, a NC licensed general contractor, a NC licensed HVAC contractor, a NC licensed Home Inspector, a registered design professional, *a certified BPI Envelope Professional or a certified HERS rater.*

During testing:

    1. Exterior windows and doors, fireplace and stove doors shall be closed, but not sealed;

    2. Dampers shall be closed, but not sealed, including exhaust, backdraft, and flue dampers;

    3. Interior doors shall be open;

    4. Exterior openings for continuous ventilation systems, air intake ducted to the return side of the conditioning system, and energy or heat recovery ventilators shall be closed and sealed;

    5. Heating and cooling system(s) shall be turned off; and

    6. Supply and return registers shall not be sealed.

The air leakage information, including building air leakage result, tester name, date, and contact information, shall be included on the certificate described in Section 401.3.

For Test Criteria 1 above, the report shall be produced in the following manner: Perform the blower door test and record the *CFM50*_____. Calculate the total square feet of surface area for the building thermal envelope, all floors, ceilings, and walls (this includes windows and doors) and record the area_____. Divide *CFM50* by the total square feet and record the result below. If the result is less than or equal to **[0.30 CFM50/SFSA]** the envelope tightness is acceptable; or

For Test Criteria 2 above, the report shall be produced in the following manner: Perform a blower door test and record the *CFM50*_____. Multiply the CFM50 by 60 minutes to create CFHour50 and record _____. Then calculate the total conditioned volume of the home and record_____. Divide the CFH50 by the total volume and record the result below. If the result is less than or equal to **[5 ACH50]** the envelope tightness is acceptable.

Property Address: _____

Fan attachment location: _____Company Name: _____

Contact Information: _____

_____   _____
Signature of Tester   Date

Permit Holder, NC Licensed General Contractor, NC Licensed HVAC Contractor, NC Licensed Home Inspector, Registered Design Professional, Certified BPI Envelope Professional, or Certified HERS Rater

(circle one)

4D.3
Duct sealing. Duct air leakage test (Section 403.2.2)
Sample Worksheet for Alternative Residential Energy Code for Higher Efficiency

403.2.2 Sealing. All ducts, air handlers, filter boxes and building cavities used as ducts shall be sealed. Joints and seams shall comply with Part V – Mechanical, Section 603.9 of the *North Carolina Residential Code*.

Duct tightness shall be verified as follows:

Total duct leakage less than or equal to 4 CFM (12 L/min) per 100 ft² (9.29 m²) of *conditioned floor area* served by that system when tested at a pressure differential of 0.1 inches w.g. (25 Pa) across the entire system, including the manufacturer's air handler enclosure.

During testing:

1. Block, if present, the ventilation air duct connected to the conditioning system.
2. The duct air leakage testing equipment shall be attached to the largest return in the system or to the air handler.
3. The filter shall be removed and the air handler power shall be turned off.
4. Supply boots or registers and return boxes or grilles shall be taped, plugged, or otherwise sealed air tight.
5. The hose for measuring the 25 Pascals of pressure differential shall be inserted into the boot of the supply that is nominally closest to the air handler.
6. Specific instructions from the duct testing equipment manufacturer shall be followed to reach duct test pressure and measure duct air leakage.

Testing shall be performed and reported by the permit holder, a NC licensed general contractor, a NC licensed HVAC contractor, a NC licensed Home Inspector, a registered design professional, a certified BPI Envelope Professional or a certified HERS rater. A single point depressurization, not temperature corrected, test is sufficient to comply with this provision, provided that the duct testing fan assembly has been certified by the manufacturer to be capable of conducting tests in accordance with ASTM E 1554-07.

The duct leakage information, including duct leakage result, tester name, date, company and contact information, shall be included on the certificate described in Section 401.3.

For the Test Criteria, the report shall be produced in the following manner: perform the HVAC system air leakage test and record the CFM25. Calculate the total square feet of Conditioned Floor Area (CFA) served by that system. Multiply CFM25 by 100, divide the result by the CFA and record the result. If the result is less than or equal to **[4 CFM25/100 SF]** the HVAC system air tightness is acceptable.

Complete one duct leakage report for each HVAC system serving the home:

Property Address: _____

HVAC System Number: _____ Describe area of home served: _____

CFM25 Total _____. Conditioned Floor Area (CFA) served by system: _____ s.f.

CFM25 x 100 divided by CFA = _____ CFM25/100SF (e.g. 100 CFM25 x 100/2,000 CFA = 5 CFM25/100SF)

Fan attachment location: _____

Company Name: _____

Contact Information: _____

_____   _____
Signature of Tester   Date

Permit Holder, NC Licensed General Contractor, NC Licensed HVAC Contractor, NC Licensed Home Inspector,
Registered Design Professional, Certified BPI Envelope Professional, or Certified HERS Rater
(circle one)

APPENDIX 5

Statement of Compliance – HVAC System Installation

Project Name: _____

Project Location: _____

In my professional opinion, the HVAC systems have been installed and are in substantial compliance with the intent of the approved project plans and specifications based on a site observation performed on_____and upon review of the following:_____*date(s)*

Yes	No	Not Required	Items	Comments
			Testing and Balance Report	
			Operations and maintenance manuals	
			HVAC Equipment	
			Control sequences	

Name: _____

Signature: _____

Date: _____

Seal: _____

INDEX

IECC TOOLS

ICC Code Resources help you learn, interpret and apply the IECC effectively

IN THE FIELD

FITS IN YOUR POCKET!
ENERGY INSPECTOR'S GUIDE: BASED ON THE 2009 INTERNATIONAL ENERGY CONSERVATION CODE® AND ASHRAE/IESNA 90.1-2007

Your ideal resource for effective, accurate, consistent, and complete commercial and residential energy provisions. This handy pocket guide is organized in a manner consistent with the inspection sequence and process for easy use on site. Increase inspection effectiveness by focusing on the most common issues relevant to energy conservation. (76 pages)

SOFT COVER #7808S09
PDF DOWNLOAD #8886P09

ENERGY EFFICIENCY CERTIFICATE STICKERS

The energy provisions in IRC® Section N1101.8 and IECC Section 401.3 require a type of certificate be installed. This easy-to-use sticker clearly lists the general insulation, window performance, and equipment efficiency details. Sold in packets of 25.
#0726S

CODE SOURCE: ENERGY CONSERVATION CODE

This value-packed resource will serve as a helpful in-the-field reference guide and as a critical component of the code enforcement and inspection process. Designed to assist field inspectors and plans examiners in the completion and performance of their duties, it will instill a solid knowledge of the practical application of the 2009 IECC and the standards set forth by the American Recovery and Reinvestment Act (ARRA).

Features:
- Lists the most common code items warranting examination for compliance.
- Organizes them in a manner that is both efficient and effective.
- Comprehensive coverage prepares building industry professionals and students for safe, accurate, and code-compliant work.
- The "quick tab" format allows easy access to critical information.
- Durable laminated pages withstand a variety of field conditions. (75 pages)
#4866S09

LEARN MORE ABOUT IECC TOOLS TODAY! 1-800-786-4452 | www.iccsafe.org/store

COMMENTARY

CODE AND COMMENTARY TOGETHER!
2009 IECC: CODE AND COMMENTARY

This convenient and informative resource contains the full text of the IECC, including tables and figures, followed by corresponding commentary at the end of each section in a single document.

· Read expert Commentary after each code section.
· Learn to apply the codes effectively.
· Understand the intent of the 2009 IECC with help from the code publisher.

The CD and download versions contain the complete Code and Commentary text in PDF. Search text, figures and tables; or copy and paste small excerpts into correspondence or reports. (245 pages)

SOFT COVER #3810S09
PDF DOWNLOAD #878P09
CD-ROM (PDF) #3810CD09

STUDY TOOLS

GREAT EXAM PREP TOOL!
2009 INTERNATIONAL ENERGY CONSERVATION CODE STUDY COMPANION

The ideal way to master the code for everyday application, to prepare for exams, or lead a training program. This Study Companion provides a comprehensive overview of the energy conservation provisions of the 2009 IECC, including the requirements for both residential and commercial energy efficiency. (188 pages)

· 10 study sessions contain key points for review, applicable code text and commentary.
· 20-question quiz at end of each study session.
· Helpful answer key lists the code reference for each question.

Great resource for Certification exams: Commercial Energy Inspector, Commercial Energy Plans Examiner, Residential Energy Inspector/Plans Examiner, or Green Building—Residential Examiner

SOFT COVER #4807S09
PDF DOWNLOAD #8787P09

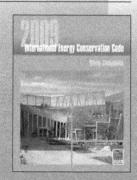

FLASH CARDS: 2009 IECC

Provides code users, students and exam candidates with an effective, time-tested, easy-to-use method for study and information retention. Prepared and reviewed by code experts to ensure accuracy and quality. (60 cards)

#1821S09
BUY THE IECC STUDY COMPANION AND FLASH CARDS TOGETHER AND SAVE!
#4807BN09

Question   Answer

LEARN MORE ABOUT IECC TOOLS TODAY! 1-800-786-4452 | www.iccsafe.org/store